CAMBRIDGE NATIONAL

LEVEL 1/LEVEL 2

T0173223

ENGINEERING DESIGN

J822

Jonathan Adams,
Peter Valentine & Alex Reynolds

Boost

HODDER
EDUCATION
AN HACHETTE UK COMPANY

The teaching content of this resource is endorsed by OCR for use with specification OCR Level 1/Level 2 Cambridge National in Engineering Design (J822).

All references to assessment, including assessment preparation and practice questions of any format/style are the publisher's interpretation of the specification and are not endorsed by OCR.

This resource was designed for use with the version of the specification available at the time of publication. However, as specifications are updated over time, there may be contradictions between the resource and the specification, therefore please use the information on the latest specification and Sample Assessment Materials at all times when ensuring students are fully prepared for their assessments.

Endorsement indicates that a resource is suitable to support delivery of an OCR specification, but it does not mean that the endorsed resource is the only suitable resource to support delivery, or that it is required or necessary to achieve the qualification.

OCR recommends that teachers consider using a range of teaching and learning resources based on their own professional judgement for their students' needs. OCR has not paid for the production of this resource, nor does OCR receive any royalties from its sale. For more information about the endorsement process, please visit the OCR website.

Although every effort has been made to ensure that website addresses are correct at time of going to press, Hodder Education cannot be held responsible for the content of any website mentioned in this book. It is sometimes possible to find a relocated web page by typing in the address of the home page for a website in the URL window of your browser.

Hachette UK's policy is to use papers that are natural, renewable and recyclable products and made from wood grown in well-managed forests and other controlled sources. The logging and manufacturing processes are expected to conform to the environmental regulations of the country of origin.

Orders: please contact Hachette UK Distribution, Hely Hutchinson Centre, Milton Road, Didcot, Oxfordshire, OX11 7HH. Telephone: +44 (0)1235 827827. Email education@hachette.co.uk Lines are open from 9 a.m. to 5 p.m., Monday to Friday. You can also order through our website: www.hoddereducation.co.uk

ISBN: 978 1 3983 5033 5

First published in 2022 by
Hodder Education,
An Hachette UK Company
Carmelite House
50 Victoria Embankment
London EC4Y 0DZ

www.hoddereducation.co.uk

Impression number   10 9 8 7 6 5 4 3

Year   2026 2025 2024 2023

Cover photo © stokkete - stock.adobe.com
Typeset in India by Integra Software Services Pvt Ltd
Produced by DZS Grafik, Printed in Slovenia
A catalogue record for this title is available from the British Library.

Contents

Introduction

This book has been designed to help you develop the knowledge, understanding and practical skills you'll need to complete the OCR Cambridge National in Engineering Design (J822) qualification. It will give you an insight into what it is like to work in the engineering design and development industry. You will learn how to use both 2D and 3D engineering techniques to communicate engineering design ideas and how to design new products to meet a design brief.

The qualification includes three units; all of these are covered in this book and you must study all three units.

R038 Principles of engineering design

In this unit you will learn about the different stages involved in the design process. You will learn about different designing requirements and how to communicate design outcomes and evaluate design ideas.

This unit is assessed by a written exam, which is set and marked by OCR. Your teacher will tell you when you will complete this exam. The exam will last one hour and 15 minutes and will have two sections – Sections A and B:

- Section A includes 10 multiple choice questions and is worth 10 marks; and
- Section B has a mixture of short answer questions and extended response questions. This section has 60 marks.

R039 Communicating designs

In this unit you will learn how to communicate your design ideas using a range of techniques, including manual production of freehand sketches and engineering drawings, as well as using computer aided design (CAD).

This unit is assessed by an assignment that includes four practical tasks you will need to complete. It will take you 10–12 hours to complete the assignment. There are 60 marks available.

R040 Design evaluation and modelling

In this unit you will learn how to model design ideas and test them as well as how to evaluate products.

This unit is assessed by an assignment that includes six practical tasks you will need to complete. It will take you 10–12 hours to complete the assignment. There are 60 marks available.

Acknowledgements

The authors and Publishers would like to thank the following schools for permission to use examples of their students' project work: NUAST (Nottingham University Academy of Science and Technology), Selby High School, St Paul's Catholic School, UTC Sheffield City Centre and WMG Academy for Young Engineers.

We would also like to thank Shania Edwards and Katie Bebbington, photography students at the University of Northampton, who took the photographs on pages 173–7, and Dr Mohammad Ghaleeh, who kindly helped with the following CAD drawings: Figures 2.36, 2.37, 2.38, 2.40, 2.41, 2.43, 2.51, 2.52, 2.53, 2.54, 2.55, 2.56, 2.57, 2.58, 2.59, 2.60, 2.61, 3.19, 3.20, p.184 and p.185.

The Publishers would like to thank the following for permission to reproduce copyright material.

Every effort has been made to trace all copyright holders, but if any have been inadvertently overlooked, the Publishers will be pleased to make the necessary arrangements at the first opportunity.

p.1 © Cat027/stock.adobe.com; Fig 1.3 © AS Photo Project/stock.adobe.com; Fig 1.5 © Pxl.store/stock.adobe.com; Fig 1.6 © Elroi/stock.adobe.com; Fig 1.7 © Daxiao Productions/stock.adobe.com; Fig 1.8 © Bernardbodo/stock.adobe.com; Fig 1.9 © Juliars/stock.adobe.com; Fig 1.10 © Gérard Bottino/stock.adobe.com; Fig 1.11 © Wingedbull/stock.adobe.com; Fig 1.12 © Chaosamran_Studio/stock.adobe.com; Fig 1.13 © DigitalGenetics/stock.adobe.com; Fig 1.14 © Steve Mann/stock.adobe.com; Fig 1.15 © pavlodargmxnet/stock.adobe.com; Fig 1.16 © Sepia100/stock.adobe.com; Fig 1.17 © 3DConcepts/stock.adobe.com; Fig 1.18 © mari1408/stock.adobe.com; Fig 1.19 © massarfabi2016/stock.adobe.com; Fig 1.20 © Chaimongkol/stock.adobe.com; Fig 1.22 © highwaystarz/stock.adobe.com; Fig 1.23 © toa555/stock.adobe.com; Fig 1.24 © Ruslan Ivantsov/Shutterstock.com; Fig 1.25 © mauvries/stock.adobe.com; Fig 1.26 © Radub85/stock.adobe.com; Fig 1.27 © Hulton Archive/Getty Images; Fig 1.28 © Starush/stock.adobe.com; Fig 1.29 © apopium/stock.adobe.com; Fig 1.30 © CG Bear/stock.adobe.com; Fig 1.31 © bergamont/stock.adobe.com; Fig 1.32 © Moose/stock.adobe.com; Fig 1.34 © Gorodenkoff/stock.adobe.com; Fig 1.35 © Juan Pablo Turén/stock.adobe.com; Fig 1.36 ©Africa Studio/stock.adobe.com; Fig 1.37 © alexlmx/stock.adobe.com; Fig 1.38 © Kv_san/stock.adobe.com; Fig 1.39 © Savo Ilic/stock.adobe.com; Fig 1.40 © Gavran333/stock.adobe.com; Fig 1.41 © antonmatveev/stock.adobe.com; Fig 1.42 © Alexandr Bognat/stock.adobe.com; Fig 1.43 © Stephen/stock.adobe.com; Fig 1.44 © Vitals/stock.adobe.com; Fig 1.45 © Sundry Photography/Shutterstock.com; Fig 1.46 © Ljupco Smokovski/stock.adobe.com; Fig 1.47 © eyeretina/stock.adobe.com; Fig 1.48 © Oyoo/stock.adobe.com; Fig 1.51 © Surasak/stock.adobe.com; Fig 1.52 © Batanin/stock.adobe.com; Fig 1.53 © Andrea/stock.adobe.com; Fig 1.54 © Andrew Harker/Shutterstock.com; Fig 1.55 Yorkshire Pics/Alamy Stock Photo; Fig 1.57 © danmorgan12/stock.adobe.com; Fig 1.58 © I Viewfinder/stock.adobe.com; Fig 1.59 © jeson/stock.adobe.com; Fig 1.63 © Stockphoto-graf/stock.adobe.com; Fig 1.64 © sNike/Shutterstock.com; Fig 1.65 © Eimantas Buzas/stock.adobe.com; Fig 1.66 © Mehmetcan/stock.adobe.com; Fig 1.67 © ikonoklast_hh/stock.adobe.com; Fig 1.74 © Kadmy/stock.adobe.com; Fig 1.75 © Vasyl/stock.adobe.com; Fig 1.76 © oyoo/stock.adobe.com; Fig 1.77 © Nataliya Hora/stock.adobe.com; Fig 1.78 © Yuri Bizgaimer/stock.adobe.com; Fig 1.79 © Ramil Gibadullin/stock.adobe.com; Fig 1.80 © Inus Grobler/stock.adobe.com; Fig 1.81 © Joe Gough/stock.adobe.com; Fig 1.82 © Andrei Kholmov/Shutterstock.com; Fig 1.84 © Greg mercurio/stock.adobe.com; Fig 1.85 © Proxima Studio/stock.adobe.com; Fig 1.86 © Urupong/stock.adobe.com; Fig 1.87 © 2022 The British Standards Institution; Fig 1.88 Crown copyright; Fig 1.89 © MigrenArt/stock.adobe.com; Fig 1.90 © Lavabereza/stock.adobe.com; Fig 1.91 © Dalibor Danilovic/stock.adobe.com; Fig 1.92 © Pavel/stock.adobe.com; Fig 1.93 © moodboard/stock.adobe.com; Fig 1.94 © markobe/stock.adobe.com; Fig 1.95 © Dmitry Zimin/stock.adobe.com; Fig 1.97 © 3DConcepts/stock.adobe.com; Fig 1.98 © Justanotherspare/stock.adobe.com; Fig 1.99 © Cherezoff/stock.adobe.com; Fig 1.109 © Vladimir Nikiforov/stock.adobe.com; Fig 1.110 © Ronstik/Alamy Stock Photo; Fig 1.111 © Gorodenkoff/stock.adobe.com; Fig 1.112 © Rdnzl/stock.adobe.com; Fig 1.113 © Coprid/stock.adobe.com; Fig 1.114 © Kitti/stock.adobe.com; Fig 1.115 © LuchschenF/stock.adobe.com; Fig 1.116 © AVD/stock.adobe.com; Fig 1.117 © Sergojpg/stock.adobe.com; Fig 1.118 © Andrzej Tokarski/stock.adobe.com; Fig 1.119 © Rawpixel.com/stock.adobe.com; Fig 1.120 © PriceM/stock.adobe.com; p.104 © Chaosamran_Studio/stock.adobe.com; Fig 2.1 © Artem Shadrin/stock.adobe.com; Fig 2.2 © Artem Shadrin/stock.adobe.com; Fig 2.5 © Stuart Douglas (Rugby School Thailand); Fig 2.6 © Stuart Douglas (Rugby School Thailand); Fig 2.10 © Stuart Douglas (Rugby School Thailand); Fig 2.13 © Stuart Douglas (Rugby School Thailand); Fig 2.14 © Stuart Douglas (Rugby School Thailand); Fig 2.19 © Zern Liew/stock.adobe.com; Fig 2.20 © Jenesesimre/stock.adobe.com; Fig 2.21 © Bioraven/stock.adobe.com; Fig 2.22 © Stuart Douglas (Rugby School Thailand); Fig 2.23 © Stuart Douglas (Rugby School Thailand); Fig 2.25 © Kaninstudio/stock.adobe.com; Fig 2.39 © NUR AZIZUL/Shutterstock.com; Fig 2.42 © Tatyana Petrova/stock.adobe.com; Fig 2.45 © Arkadivna/Shutterstock.com; Fig 2.46 © Africa Studio/stock.adobe.com; Fig 2.62 © Mathew/stock.adobe.com; Fig 2.63 © Lalandrew/stock.adobe.com; Fig 2.64 © 2002lubava1981/stock.adobe.com; p.150 © Mathew/stock.adobe.com; Fig 3.1 © Kaspars Grinvalds/stock.adobe.com; Fig 3.2 © deagreez/stock.adobe.com; Fig 3.3 © thaiprayboy/stock.adobe.com; Fig 3.4 © Scanrail/stock.adobe.com; Fig 3.7 © Africa Studio/stock.adobe.com; Bike Light 1 © BY-_-BY/stock.adobe.com; Bike Light 2 © Maksim_e/stock.adobe.com; Bike Light 3 © Oleg/stock.adobe.com; Fig 3.8 This illustration used under license from J H Haynes & Co Ltd.; Fig 3.9 © taesmileland/123RF; Fig 3.10a © 73kPx/stock.adobe.com, b © complize | M.Martins/stock.adobe.com, c © RolandoMayo/stock.adobe.com, d © mewaji/stock.adobe.com; Fig 3.11 © Antonioguillem/stock.adobe.com; Fig 3.12 © Caliber/Shutterstock.com; Fig 3.13 © 2002lubava1981/stock.adobe.com; Fig 3.14 © Photoplotnikov/stock.adobe.com; Fig 3.15 © Hoomoo/stock.adobe.com; Fig 3.16 © Bert Folsom/stock.adobe.com; Fig 3.18 © Africa Studio/stock.adobe.com; Fig 3.21 © Tan Kian Khoon/stock.adobe.com; Fig 3.22 © ilyukov/stock.adobe.com; Fig 3.29 © Pixel_B/stock.adobe.com; Fig 3.32 © Maskalin/stock.adobe.com; Fig 3.33a © Yuri Bizgaimer/stock.adobe.com, b © aquatarkus/stock.adobe.com; Fig 3.42 © Yuthayut Chanthabutr/stock.adobe.com; Fig 3.43 hard hat © kitthanes/stock.adobe.com, ear defenders © Igor Sokolov/stock.adobe.com, work coat © mihailgrey/stock.adobe.com, mask © phanasitti/stock.adobe.com, goggles © showcake/stock.adobe.com; Fig 3.45 © Stokkete/stock.adobe.com; Fig 3.46 © Madarakis/stock.adobe.com; Fig 3.47 © Goodcat/Shutterstock.com; Fig 3.48 © Kirill4mula/stock.adobe.com; Fig 3.49 © tanantornanutra/stock.adobe.com; Fig 3.50 © Kenjo/stock.adobe.com; Fig 3.56 © 3dmitruk/stock.adobe.com; Fig 3.58 © yanik88/stock.adobe.com; Fig 3.59 © warut/stock.adobe.com; Fig 3.60 © oyoo/stock.adobe.com

All other photos not credited were supplied by the authors.

How to use this book

This textbook contains all three units for the redeveloped Cambridge National Engineering Design Level 1/Level 2 qualification (J822).

These units are:

- Unit R038 Principles of engineering design
- Unit R039 Communicating designs
- Unit R040 Design evaluation and modelling

Each unit is divided into topic areas. All of the teaching content for each topic area is covered in the book.

Key features of the book

A range of learning activities are included in the Student Book. They can be used flexibly to embed and supplement learning.

Each chapter (or unit) begins with flexible reference material. This content can be used to introduce the unit, the topics covered and the method of assessment. It will also help to empower the students to take control of their own revision and assessment preparation.

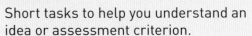

Topic areas

A clear statement of the topic areas so you know exactly what is covered.

How will I be assessed?

Assessment methods are clearly listed and fully mapped to the specification.

There is also a range of in-chapter learning features to support your teaching.

Each of these learning features is showcased in the sample chapter, so you can consider how you will use them in the classroom and with your students.

Getting started

Short activities to introduce you to the topic.

Key terms

Definitions of important terms.

Activities

Short tasks to help you understand an idea or assessment criterion.

Case study

Real-life scenarios to show how concepts can be applied to businesses.

Research

Activities that draw on the content covered in the book, to reinforce understanding.

Test your knowledge

Questions to test your knowledge and understanding of each learning outcome. Answers can be found online at: www.hoddereducation.co.uk/cambridge-nationals-2022/answers

Synoptic links

Links to relevant details in other parts of the book so you can see how topics link together.

Practice questions

This feature appears in Unit R038, which will be assessed via an exam. It includes practice questions, mark schemes and example answers to help you prepare for the exam. Answers can be found online at: www.hoddereducation.co.uk/cambridge-nationals-2022/answers

Assignment practice

This feature appears in other units and will help you prepare for non-examined assessment with model assignments, mark schemes and tips.

Unit R038

Principles of engineering design

About this unit

This unit will help you to understand different design strategies and the stages involved in designing products. You will discover what information is needed to produce design briefs and design specifications. You will focus on iterative design, which is one of the most widely used design strategies, and look at other design influences, including market demand, sustainability and manufacturing. Finally, you will learn how designers communicate their ideas using engineering drawings, and virtual and physical prototypes.

Topic areas

In this unit, you will learn about:

1 Designing processes
2 Design requirements
3 Communicating design outcomes
4 Evaluating design ideas

Topic area 1 Designing processes

Getting started

Think about a product that you use every day. This could be something complex, like your laptop or mobile phone, or a simpler product, such as a plastic bottle or your toothbrush.

In pairs, discuss some of the key product features that would have been added to the design based on the needs of the user.

1.1 The stages involved in design strategies

There are many different **design strategies**. A designer chooses the most appropriate strategy for the product being designed. Design strategies follow a **design process**. Designers need to go through a series of stages as part of the design process to enable products to be created successfully. The design process needs to be structured and organised to allow for efficient design of the product.

Linear design

The **linear design** process involves a series of stages that are carried out one after the other, as shown in Figure 1.1. There can be more or fewer stages in the process, but the basic stages to get to a final design are: exploring the problem, researching, generating design ideas, producing models, testing, evaluating the design and making final modifications. In linear design, each step must be completed fully before moving on to the next step. This process allows designers to correct some design errors, but usually they cannot go back to a previous stage once finished. This means that any errors in the design might not be corrected, so the product might not always be the best design solution. However, it is one of the simplest and quickest design strategies.

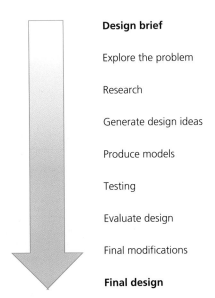

Figure 1.1 Linear design

Key terms

Design strategy A series of stages that are part of the design process.

Design process A series of stages that designers and engineers use in creating functional products.

Linear design A design process where the stages are carried out one after another, often without turning to any of the previous stages.

Iterative design

Unlike linear design, in the **iterative design** process, each of the stages (sometimes called phases) is revisited and reflected upon regularly to improve and refine design ideas. This ensures that the product meets the needs of the **user** and satisfies the design brief. It also makes sure that the requirements of the **client** (who commissioned the design) are met. If manufacturing is also considered as part of the process, then it is also possible to make sure the product can be economically and sustainably manufactured.

Iterative design is a circular design process that models, evaluates and improves designs based on the results of testing. An example of the iterative **design cycle** is shown in Figure 1.2.

A typical iterative design cycle includes a series of phases:

- Identify – analysing the design brief, carrying out research and planning
- Design – generating design ideas, including sketches and drawings, writing a design specification
- Optimise – producing prototypes and making design improvements
- Validate – testing designs, reviewing design decisions and comparing final design against design brief and design specification.

Key terms

Iterative design A circular design process that models, evaluates and improves designs based on the results of testing.

User Person or people who will use the final product.

Client Person, group of people or company that has commissioned the development of a new product.

Design cycle A set of processes, split into four phases, that designers follow to ensure efficient and effective product development.

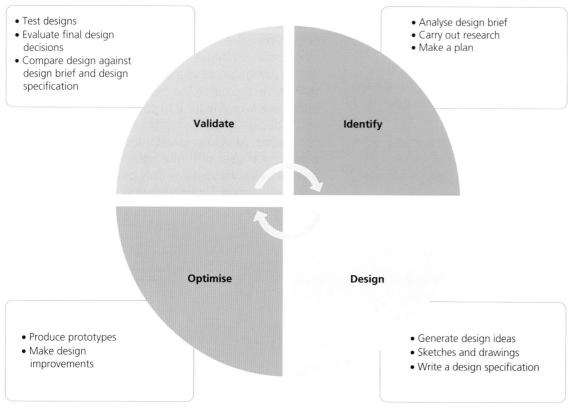

- Test designs
- Evaluate final design decisions
- Compare design against design brief and design specification

Validate

- Analyse design brief
- Carry out research
- Make a plan

Identify

Optimise

- Produce prototypes
- Make design improvements

Design

- Generate design ideas
- Sketches and drawings
- Write a design specification

Figure 1.2 Iterative design

As the process is iterative, the designer makes modifications to the design throughout the design process. However, the complete design process might take longer than linear design depending on how many times the designer needs to go round the cycle. At some point, the designer will need to complete the design due to the time and **budget** allowed for designing the product. The iterative design process will be covered in more detail later in this unit.

Inclusive design

Inclusive design is a design process in which a product is optimised for a specific user with specific needs. Sometimes, this user is called an extreme user, which means that they have specific needs that are often overlooked with other design processes. An example is the wheelchair user in Figure 1.3, who will have different needs for products to accommodate their different abilities. Designing products particularly for their requirements is an example of inclusive design. This might be more time-consuming to get the design right, but there are often many benefits to the user. Highly specialised products might also have a limited number of potential users, so the cost to manufacture those products and the selling price might be higher due to low production quantities. However, taking an inclusive design approach often has advantages for other users too. For example, adding ramps for wheelchair users in public buildings makes access easier for other user groups, such as parents with pushchairs and site maintenance staff.

Figure 1.3 Wheelchair users require inclusive design

User-centred design

User-centred design is another example of an iterative design process in which the designer focuses on the user and their needs at each step of the design process. It involves consulting with users throughout the design process using a variety of research methods and design techniques, to create highly usable and accessible products. Figure 1.4 shows a typical user-centred design process. It is important to understand clearly where and how the product will be used and the specific requirements of the user throughout the design process. As it is an iterative process, the designer will often return to these requirements throughout the design of a product to make refinements and improvements. Advantages of this design process include creating products that perfectly match the needs of the target user. However, the time to complete all stages of the design might be longer than other design processes. Also, a product specially designed for a particular user or limited group of users might be more expensive if it is only manufactured in small quantities. Examples of highly user-centred products are smart watches and computer apps.

Key terms

Budget Amount of money allocated by a client or company to develop a product.

Inclusive design A design process where a product is optimised for a specific user with specific needs.

User-centred design A design process where the designer focuses on the user and their needs in each step of the design process.

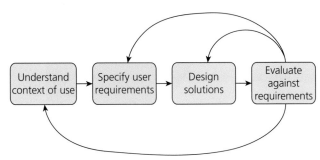
Figure 1.4 User-centred design

Sustainable design

Sustainable design is a design process that aims to reduce the negative impacts of a product on the environment. The main objectives of sustainable design are to reduce the use of non-renewable resources, to minimise waste and to create products that are manufactured and operate using less energy. This can be achieved by designing products that use fewer precious resources in their manufacture and operation, and products that can be reused, repaired and easily recycled. When designing products with **sustainability** in mind, the designer might have to make compromises to the design, such as selecting different materials and making the product function in less optimal ways. **Sustainable** products might be more expensive due to the technology and manufacturing techniques required to produce them. An electric car (Figure 1.5) is an example of a product designed for sustainability which is appealing to customers but that is also relatively expensive to buy due to the lightweight materials and battery technology required. However, consumers often prefer products that have been designed for sustainability, which can improve the reputation of the manufacturer and increase sales.

Figure 1.5 Electric cars are an example of a sustainable product

Ergonomic design

Ergonomic design is the process of designing products so that they perfectly fit the people who use them. It aims to improve products to minimise the risk of injury or harm, and to make them easier to use. **Ergonomics** uses **anthropometric** data (which we will learn more

about later in this unit) to determine the best size, shape and form of a product. The office chair shown in Figure 1.6 has been designed to fit the user perfectly and to be easy to adjust so that it reduces the risk of being harmed when in use. It is an example of ergonomic design. To carry out ergonomic design, the designer needs to know the target user and data about them, so the design process can take longer. Also, if the product is designed for a particular group of users, then this might limit its selling potential and the selling price of the product might need to increase due to the reduced quantity being manufactured – for example, a pair of scissors designed for left-handed users would only appeal to approximately 10 per cent of the population.

Figure 1.6 Ergonomic office chair

Key terms

Sustainable design A design process where the designer attempts to reduce negative impacts of a product on the environment.

Sustainability Meeting current needs without preventing future generations from meeting their needs.

Sustainable When something is used in a way that ensures it does not run out.

Ergonomic design A process for designing products using anthropometric data so that they perfectly fit the people who use them.

Ergonomics Science of designing products so that users can interact with them as efficiently and comfortably as possible.

Anthropometrics Study of the measurements of the human body.

Activity

Draw a table and summarise the key characteristics, advantages and disadvantages of each of the different design strategies. The first one has been done for you.

	Key characteristics	Advantages	Disadvantages
Linear design	Step-by-step process, moving from one step to the next ● Each step must be completed before moving onto the next ● Cannot usually go back to previous stages	● Perhaps the quickest and simplest design process	● Errors in design cannot always be corrected so the final design might not be the best
Iterative design			
Inclusive design			
User-centred design			
Sustainable design			
Ergonomic design			

Test your knowledge ✔

What are some of the characteristics of the following design processes?

- linear design
- iterative design
- inclusive design
- user-centred design
- sustainable design
- ergonomic design.

1.2 Stages of the iterative design process

As we saw in the previous section, designers often use an iterative design process involving a design cycle to give structure and sequence to the process of creating a new product. By defining this structure, designers can ensure they have all the information they need at each stage of the cycle to avoid costly and time-consuming changes to the design.

Key term

Target market Group of people at whom the product being developed is aimed.

The design cycle (Figure 1.2) has four phases:

- identify phase
- design phase
- optimise phase
- validate phase.

At the completion of each phase, the designer can review the development of the product to ensure it meets the design brief, is in line with client and user requirements and is still on track to be completed on time and within budget. The designer might make several cycles of the design process, or move back and forward between phases before arriving at the final design.

If designers do not follow such an organised process, they risk spending large amounts of time developing a product that does not meet the expectations of the client, user or **target market**. They may then have to return to earlier stages of the cycle to make changes, which would make the design process less efficient and result in delays. In extreme cases, designers and businesses may even launch products that are not successful, which can negatively affect sales and the company's reputation.

Design

Identify phase

The first phase of the design cycle:

- ensures the designer has a clear understanding of the requirements of the design brief
- defines client and user needs by carrying out research
- considers the processes to be followed throughout the development of the design (process planning).

For the designer, this phase confirms client and user expectations of the completed product and gives them an understanding of the market and which competitor products already exist. It also allows the designer to define the scope of the project, taking into consideration both design and **manufacturing processes** as well as the costs of product development, to ensure it can be completed on time and within budget.

Analysis of the design brief

A design brief is either supplied by the client or developed as a collaboration between the client and the designer. It sets out requirements for the product that is going to be designed. It also describes the problem that needs to be solved and includes the main features or characteristics that the product must have to be successful.

The designer needs to think through the requirements of the brief to ensure they fully understand:

- what the product must do
- who will use the product
- what the client expects from the product.

It is important that the designer and the client agree on the requirements of the brief before conducting research and starting to design the product. This allows the designer to research appropriate areas, such as the target users and product needs. If these requirements are not clearly defined, the designer could easily begin to develop a product that is not fit for purpose.

Figure 1.7 A designer working with a client

Carrying out research

Carrying out research is an important part of the initial phase of the design cycle. It allows the designer to explore and confirm the needs of the client and the user, and therefore ensure the product meets those requirements.

If a designer produced a design without first researching client and user needs, they could make decisions about the product that were incorrect. This could result in expensive product development requiring large changes, or even a product that was unsuccessful when put on sale. Research is also important so the designer can find out what the product needs in order to function as required.

First, the designer should consider issues such as where the product will be used (the **working environment**) and how the user will hold or interact with the product (ergonomics).

Key terms

Manufacturing process Stages through which raw materials go in order to be transformed into a product.

Working environment Place where a product will be used or situated during operation.

They could then research what materials are available, what **components** could be used or what products currently exist on the market that offer similar solutions.

By carrying out this research early in the design cycle, it is possible to make informed decisions about the design of the product, such as how the **geometry** of the product will allow certain materials or manufacturing processes to be used.

Types of research

There are two main types of research that can be carried out by the designer: primary and secondary. These are covered in more detail, and you will perform your own product research activities, in Unit R040.

Primary research is original research that is conducted first hand, so in this case the designer would conduct the research themselves. Examples include:

● questionnaires
● surveys
● focus groups
● interviews.

The advantage of primary research is that it provides current, up-to-date information that is specific to the designer or company, and this can give the company an advantage over its competitors. However, it can also be time-consuming to complete, and if the number of people questioned, surveyed or interviewed isn't large enough, the results can be misleading (biased).

Secondary research is where information is gathered from sources that already exist. Examples include:

● books/magazines
● the internet
● published statistics
● existing products
● data sheets – for example, material information.

The advantage of secondary research is that it can be gathered easily, quickly and cheaply compared with primary research. However, it is unlikely to provide information that is specific to the designer's client or user. It may also be out of date, and because it often uses freely available information, it may not provide a competitive advantage.

The design brief will be based on a wide range of information from many different sources. The initial problem may be identified by the client or could be the result of consumer feedback.

To define the brief in greater detail, the designer and client may gather a range of different information through various research methods and sources.

Key terms

Components Parts or elements of a larger assembly.

Geometry Shape of an object.

Primary research Gathering original information first hand – for example, carrying out interviews, experiments, questionnaires and surveys.

Secondary research Gathering information from sources that already exist – for example, using books, newspapers, magazines and the internet.

Figure 1.8 Primary research can be carried out in a focus group

Market research

A good starting point for designers to gather data is to carry out **market research**. The client will generally have an idea of the target market when they begin initial discussions with the designer. This is important because it will help the designer to target their research at the appropriate market.

The designer may gather information about the market by using **focus groups** or distributing **surveys**.

Focus groups

A focus group is made up of people who are potential customers for a new product. They are invited to discuss their views and opinions on their wants, needs and preferences for the product. They may be given **prototypes** to review, asked for their thoughts on existing products or asked about what they would like to see included in future products released by the company.

Participants are selected because their profile (personal information) shows that they would be potential customers. However, it is important that the focus group does not become a stereotype of the perceived target market. Stereotypes can be misleading and not reflect the true nature of the target market. If the range of people in the group is too narrow, this may lead to fewer new ideas and suggestions and could limit the scope of the future market. At the same time, if the group is too broad, opinions gathered may not reflect the actual needs of future users.

Although using a focus group sounds quite simple, it can be difficult to ensure the group is made up of appropriate people and that it is inclusive and provides a range of opinions. It may be useful for a designer to use several different focus groups to ensure they gain a range of views.

Surveys

A survey is another method of gathering information from potential customers within the target market to gain their views and opinions on a new product.

Key terms

Market research Process of gathering information about the needs and preferences of potential customers.

Focus group Group of people invited to discuss their views and opinions on their wants, needs and preferences for a new product.

Survey Tool used to gather data from particular groups of people that will inform the direction of a design.

Prototype Model of a component or product created either in a software package or physically that can be used to test or check the design.

Surveys generally take the form of a questionnaire. They may be completed face to face or in written format, either on paper or online. Surveys can be distributed and repeated easily and therefore they are considered a cost-effective way of gathering information.

Surveys can focus on key groups of individuals or be more general to gain a wider view of what the general population thinks. They can also vary in duration. Some surveys are carried out over a long period of time to gain a large amount of data. The timescale of the survey depends on the development time for the product. The data gathered can then be used to make decisions about the design of the new product.

As with a focus group, it is important to ensure the information gathered is relevant to the product being developed and represents appropriate views from the potential target market.

Activity

Create a survey to gather people's views on a new design for one of the following products:

● a console game controller
● a desk chair
● a hairdryer.

Consider the type of people you will ask and what questions you would include to gain accurate information to inform the design.

The results of focus groups and surveys, alongside other information, can be used to define the needs of the target market. For example, the target market may require a low-cost solution, and this will have an impact on the manufacturing processes or materials that are used.

Changing consumer trends

A consumer **trend** is a pattern that occurs in the sales of particular products or in the behaviour of consumers. A product that is currently in trend will be extremely popular on the market and so generate high sales. For example, the introduction of wearable technology and smart watches has been a popular consumer trend, with large numbers of businesses releasing variations onto the smart-watch market.

Consumer trends change and evolve and can last for different periods of time. It is important that designers and companies understand current consumer trends and how they may change in the future. If a new consumer trend can be predicted, companies can take advantage of it by developing and releasing new products at the best time to ensure maximum sales.

> **Key term**
> **Trend** Pattern of change that can be used to predict how the demands of a market are developing.

Figure 1.9 Smart watches are an example of a consumer trend

> **Activity**
> In small groups, think of products that have been part of a consumer trend.
> - Is the product still popular?
> - What trends are appearing in the market now?

Ergonomics and anthropometrics

Ergonomics is the study of how people interact with the objects they use and the environments they use them in. Designers will often use tables of data (called anthropometric data) showing measurements of the typical human body when designing to ensure that products are the correct size for the user they are being designed for. Ergonomics and anthropometrics are covered in more detail in Topic area 2 of this unit.

Figure 1.10 Anthropometric data can be used when designing helmets

Analysis of existing products

Many products released onto the market are not completely new inventions but instead variations of products that already exist. For example, several companies have smart phones on the market with similar features.

Designers often look at competitors' products to gain a greater understanding of what is currently selling well, what features they include and where there are opportunities for improvements. They can then try to create an improved product and gain a competitive advantage when it launches onto the market.

Analysis of competitor designs may take place so that strengths and weaknesses can be identified. This analysis may look at the **aesthetics** of a product as well as how it functions, or it may include an analysis of the manufacturing process to see where improvements can be made or costs can be saved. Designers may use focus groups or surveys that focus on competitors' products.

ACCESS FM

Designers will often use tools to analyse existing products and to carry out research when designing new products. One commonly used tool is **ACCESS FM**. This stands for **A**esthetics, **C**ost, **C**ustomer, **E**nvironment, **S**ize, **S**afety, **F**unction and **M**aterials. Using these headings, and a series of questions, it is possible to research products systematically and to compare the relative strengths and weaknesses across a range of similar products.

Product disassembly

Designers will often consider how existing products have been manufactured and the materials they are made from and use this information to inform their own designs. They may carry out a **product disassembly** – this involves taking the product apart to examine the materials, parts and components that have been used to produce it.

Product analysis using techniques such as ACCESS FM and safe product disassembly is covered in more detail in Unit R040.

Key terms

Aesthetics How well a product appeals to the senses.

ACCESS FM Acronym for Aesthetics, Cost, Customer, Environment, Size, Safety, Function and Materials; a product analysis tool.

Product disassembly Taking a product apart to look at the materials, parts, components and fixings that have been used.

Figure 1.11 Product disassembly

Activity

In pairs, consider a product that you both use on a regular basis – for example, a smart phone, laptop or even a rucksack or bag. Discuss what you think are the strengths and weaknesses of your product against those of your partner's:

- Do your products share similar strengths?
- Does one product have strengths that are weaknesses in the other?
- How might you address the weaknesses while maintaining the strengths in a new design?
- Are there other products on the market that have different strengths or weaknesses?

Process planning

In order to create a successful final product on time and within budget, a designer needs to set out the processes they will follow throughout development of the design – this is called **process planning**.

For example, they need to consider the manufacturing processes that could be used. This needs to happen early in the design cycle, so that the designer can work out how long the design will take to complete and manufacture. It is also essential that the manufacturing process is considered before the physical design of the product takes place because specific manufacturing processes require certain component geometry in order to work. If a designer has not decided which manufacturing process to use, they may design components or products that cannot be manufactured.

A designer may also consider how they will prototype the product, what testing they may require and how the product would need to be maintained.

By setting out all these processes, a designer can create a timeline for production of the design, from initial concept through to final manufacture, allowing them to define the timescale (how long it should take) and calculate costs. This will result in a project plan that the designer, manufacturer and client can use to monitor progress against the timeline. It will also include checkpoints at various stages of the plan, so that the product can be reviewed against targets, such as budget and timescale for delivery.

Key terms

Process planning Where all activities required to complete the development of a new product are defined with timescales assigned, to ensure the product can be delivered on time and within budget.

Design specification Detailed document that defines all the criteria required for a new product.

Concept sketching Producing drawings quickly and often by hand in order to explore initial design ideas.

Design phase

The second phase of the design cycle is the design phase, where the designer:

- uses all the information they have gathered to create a design specification
- carries out the physical design of the product
- creates detailed engineering drawings and manufacturing plans for the product.

Production of an engineering design specification

Once the designer has gained a clear understanding of the design brief and has carried out research to define client and user needs, they then use the information they have gathered to produce an engineering **design specification**.

A design specification is a detailed document that contains all the information about the product being developed. It is created at this stage because it provides information about the physical design of the product. It provides a set of guidelines that the designer can follow and refer back to throughout the design process, to ensure the design is in line with these. It is usually presented as a list of factors that must, should or could be met.

The criteria included in a design specification are covered in Topic area 2 of this unit.

Generating a range of design ideas

There are usually many ways to solve a design problem. For example, look at the variation of designs in mobile phones, cars and sports shoes. The specifications for the products may have been very similar, but the materials and aesthetics of the final products can vary hugely.

A designer generates a range of initial design ideas that explore lots of different solutions to the design challenges and possible variations to the design. This is called **concept sketching**. The designer shares these different concepts with the client or gets feedback from the user, rather than spending a long time developing a detailed design and then finding out that the client wants to make changes. This makes the process of developing the design more efficient by avoiding costly and time-consuming modifications later on.

In order to generate these design ideas quickly, a designer will tend to use freehand 2D and 3D sketching. Sketches:

- are quick, easy and cost-effective to produce and modify
- can easily be shared with the client and user to gain feedback on the design
- do not require expensive software packages or time-consuming 3D modelling processes.

Once there is a chosen design, designers may begin to use software packages to refine it. These may include graphical sketching packages or more engineering-focused 3D **computer-aided design (CAD)** packages. At this stage, the development of the design should focus more on producing a visually accurate representation of the product, or on working out how a **mechanism** or particular function of the product works. Detailed design work happens later in the development process. See Unit R039 for more details on sketching and presenting design ideas.

Figure 1.12 Example of concept sketching

Key terms

Computer-aided design (CAD) Using computer software to develop designs for new products or components.

Mechanism Set of components that work together in a product to carry out a function.

Activity

Develop some concept sketches for new designs of the following products:

- a sports water bottle
- a smart watch
- a computer mouse.

Try using different techniques for your sketches – for example, using pencil and then pen, or adding colour. Include annotations to explain the key functions or features of your designs.

Selection and justification of chosen designs

Having generated a range of design ideas, the designer can consult the client to select those they prefer and those that are most suitable based on the specification, timescale and budget.

A designer will consider a wide range of criteria to justify which ideas are selected for more detailed development:

- In consumer products, aesthetics is usually important to ensure successful sales, so this may be one of the main reasons a design is selected for further development.
- In an engineering context – for example, automotive or aerospace components – the main factor may be the cost of the manufacturing process or the performance of the chosen design, such as strength or weight.

Figure 1.13 Examples of consumer products

Figure 1.14 Examples of engineering components

Presentation of chosen designs

The way in which a design is presented to a client should help them to understand how a product will function and what it will look like.

The presentation of a final design depends on what a product is being designed for:

- A new design of kettle will be sold to consumers, and therefore the aesthetics of the product will be an important part of the presentation to the client, alongside its function or features.
- A new design of turbine blade for a jet engine will need to show its functional advantages over previous designs (for example, increased strength, lighter weight or improved efficiency), but its appearance is less important.

In both cases, more detailed 3D CAD models would be produced at this stage:

- For presentations on aesthetics, this allows the designer to produce **rendered imagery** that showcases the product with accurate materials, finish and lighting.
- For engineering solutions, this allows the designer to work out the key characteristics of the component or product (for example, weight or volume) which can be used to inform calculations about performance.

In addition, where products are made of multiple components, initial **assembly** models can be produced that show how all the components fit

together. These models can be presented to the client as an **exploded assembly** design, which demonstrates how all the components fit together (see Unit R039, Topic areas 2 and 3 for more detail).

Development of planning and engineering drawings

At all stages of the design process, it is essential to consider how the component or product will be manufactured.

For a design to be engineered and manufactured correctly, the size and shape of the components need to be designed so that they are suitable for the chosen manufacturing process (which is covered in more detail in Unit R039, Topic area 2). Once the overall design has been finalised and approved by the client, the designer will begin to produce the final **engineering drawing** for each component, ensuring that each component has the appropriate geometry, dimensions and features.

The design cycle will often happen simultaneously, or move backwards and forwards between phases as the designer modifies a design following the results of prototyping and testing (see below), making iterative improvements to the existing design.

Key terms

Rendered imagery Photorealistic computer-generated images of products.

Assembly Putting together components to make a completed product (if the resulting product is to be incorporated into a larger product, it is called sub-assembly).

Exploded assembly A drawing where the components of a product are drawn slightly separated from each other and suspended in space to show their relationship or the order of assembly; also known as an exploded view.

Engineering drawing Type of technical drawing that details the geometry, dimensions and features of a component or product.

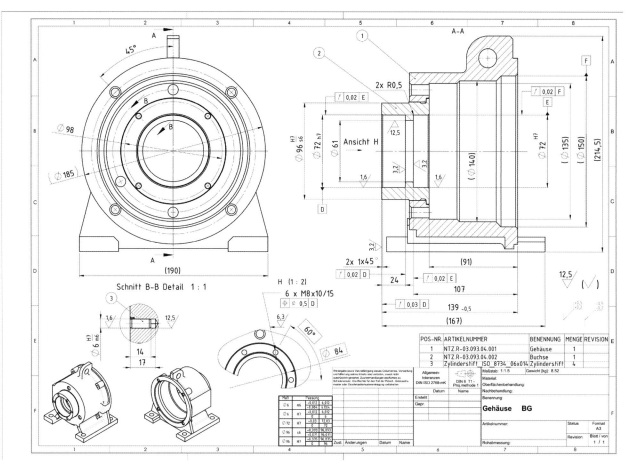

Figure 1.15 Engineering drawing

Manufacturing plans

Alongside engineering drawings, the designer will work with the manufacturer to develop plans for how the components and product will be produced.

To maximise the efficiency and cost-effectiveness of manufacture, a designer should consider this early in the design process. During market research, the designer will have analysed the potential market to assess how many products are needed. This **scale of production** will have an impact on which process is most suitable. For example, injection moulding has a large upfront cost for **tooling** and equipment but it then enables large quantities to be made at a relatively low cost per unit. This type of decision can have a big impact on a design.

Getting components or products to a position where they are ready to be manufactured is also

a time-consuming process. It includes setting up machinery, acquiring materials and creating tooling such as mould tools or **jigs and fixtures**. The longer this process takes, the longer it takes the product to reach the market.

Key terms

Scale of production Number of products to be produced to meet demand or by a certain production process – for example, one-off, batch or mass production.

Tooling Manufacturing equipment needed to produce a component, such as cutting tools, dies, gauges, moulds or patterns.

Jigs and fixtures Tools used in manufacturing to ensure components are placed or held accurately so that they can be replicated consistently.

Manufacturing plans are detailed documents that set out the materials needed, production quantity, production setup and process, and timescales. In some cases, the designer may work for a company that is able to manufacture in house, but other times the designer will have to deal with a wide range of external suppliers and contractors. Even when companies produce components in house, they still have material or machine suppliers, who are critical to the success of manufacturing.

By considering manufacturing early in the design cycle, companies can optimise components and products for manufacture and therefore reduce the time taken to develop new products and get them onto the market before their competitors. However, the company and the designer need to be sure that the design is correct before they commit to a certain method of manufacturing. If it is not, they could waste a lot of money. For example, injection-mould tools can cost tens of thousands of pounds. If design modifications are required after the mould tool is produced, the company will have invested a large amount of money in a tool that cannot be used. This is where the optimise stage of the design cycle is used to test and improve the design, to ensure it is ready for large-scale or final manufacture. Modifications cost both time and money, so it is important to ensure designs are correct at the earliest opportunity.

The design, optimise and validate phases of the design cycle will often happen simultaneously, or as a minimum, the cycle will move backwards and forwards between these phases as the designer modifies the design following the results of prototyping and testing.

Activity

In pairs, discuss the risks of developing manufacturing plans before the design has been optimised in the next phase of the design cycle. Make a list and share it with the rest of your class.

Key terms

Manufacturing plans Detailed documents that set out the material requirement, production quantity, production setup and process, and timescales for making a product.

Test of proportions Checking that the relationship between the size of different parts of a product are correct or attractive.

Test of scale (product) Checking that the overall dimensions of the product are correct or attractive.

Test of function Checking that the product works or operates in a proper or particular way.

Make and evaluate

Modelling/producing a prototype

Once the designer has generated a range of designs and selected ones that they feel are the most suitable, they can start making them. They do this through modelling and creating a prototype of a component or product, which can be virtual and/or physical. Designers usually do this in the optimise and validate phases of the design cycle to check that the product looks and feels right (to test **proportions** and **scale**, for example) and to confirm that the product operates and performs as required (to test **function**).

Optimise phase

The optimise phase of the design cycle is about finding ways that the proposed design can be improved to ensure it is the best solution to the problem.

To do this, the designer makes a prototype of the design – a model of a component or product created either in a software package or physically, allowing them to test and improve the design. The design can also be shown to clients and users to get feedback.

The process of producing a prototype, testing or gaining feedback from clients/users and then modifying the design is an iterative process (a cycle that is repeated with the aim to make improvements each time). The product may go through many different iterations to continually improve elements of the design. In some cases, this may include hundreds of different prototypes and design changes, with gradual, measurable improvement made each time to optimise the design.

Optimisation can focus on many areas, such as improving the ease of manufacture of the product, performance or ergonomics. For example, a racing car wing will go through multiple iterations to ensure it generates the best possible downforce. Designers will make numerous prototypes of the wing, in order to have multiple variations of the design to test. They will make minor modifications to the design, produce new prototypes and test the new solution until they have the optimum design.

Figure 1.16 A racing car wing will go through multiple iterations to perfect the design

When carrying out design optimisation, compromises usually need to be made. Many product development projects have a limited amount of time, money and resources to bring a new product to market. In some cases, where only minimal compromise can be made due to

product safety, efficient optimisation is critical to reduce any delays to the product development process.

The modifications may vary at different stages of the design cycle. For example, in later stages of development before going into production, changes usually focus on improvements to the final design rather than radical design modifications to multiple variations of a product.

During the optimise phase of the design cycle, designers make multiple prototypes so that they can test variations of the final design during the validate phase. These prototypes can be produced either physically or virtually.

Virtual modelling of design ideas

All physical prototyping methods are time-consuming and require the use of physical materials and resources. Processes such as **additive manufacturing** or **computer numerical control (CNC) machining** require expensive machinery and, in some cases, costly materials. Producing **virtual prototypes** gives designers the opportunity to check their designs within a computer system.

Key terms

Additive manufacturing Technologies that produce 3D components and products from CAD data by adding material layer by layer (as opposed to subtractive manufacturing processes that remove material from a larger block).

Computer numerical control (CNC) machining Using computer-controlled machine tools to remove material from a workpiece to create components.

Virtual prototype Model of a component or product produced in a software package that can be tested or used in simulations without the need to produce an actual model.

There are several benefits of producing virtual prototypes using CAD software packages:

- Virtual prototypes provide the designer with an exact, virtual replica of the product, so that they can easily confirm that its components fit together correctly and check key design factors such as weight and dimensions.
- The designer can make modifications to the design much more quickly than they could to a physical prototype.
- Virtual prototypes can be shared easily across the world, so the designer can gain rapid feedback on key features.
- Virtual prototypes can be used directly in computer-based simulations of real-world scenarios during testing, so that the performance of the design can be reviewed.

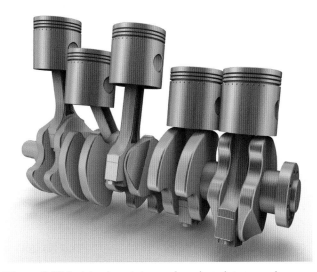

Figure 1.17 A virtual prototype of engine pistons and crankshaft assembly

Physical modelling of design ideas

Physical prototypes can be produced for many reasons and using many different techniques. In industries such as software development, it may not always be necessary to produce a physical prototype, as **simulations** can be conducted solely on a computer. However, where a product will exist physically in the real world, designers and companies will want to produce a physical prototype.

A physical prototype may simply be produced so that the aesthetics of the product can be reviewed, and feedback gained from the client or user before production begins. In other cases, the product may be made from multiple components and physical prototypes are useful to make sure those components fit together and any mechanisms work correctly.

Products that humans interact with almost always need physical prototypes. In some cases, the physical prototype will be produced so that a person can pick it up, use it and give feedback to the designer. The designer can then make ergonomic changes, such as modifying the shape of certain features, or changing the weight to make it more comfortable to use.

In addition to prototypes that check fit, form and function, designers may need to produce prototypes that are exact replicas of the final production component or product. They may need to be manufactured from the exact material, so that they share the mechanical properties of the final product. This means they can be tested under real-world conditions later in the design cycle, so that the designer knows the product can withstand the stresses on the components and is therefore safe to be put on sale.

There are many methods that can be used to produce physical prototypes. These range from manual methods to automated processes and the designer will decide on the most suitable form for the application they require. For example, to check an initial concept and ensure it is a suitable size, designers may make rough

Key terms

Physical prototype 3D physical model of a component or product that allows the designer, client or user to interact with it.

Simulation Where a computer-generated model of a component or product is exposed to virtual conditions that represent real-world scenarios to analyse how the product or component reacts.

card or foam models. These types of model are generally more useful at earlier stages of the design cycle. However, they can still be a cost-effective way of checking elements of a design at the optimise phase, due to the speed with which they can be changed and with very little material cost. Increasingly, designers have the option of producing designs directly from CAD data using additive-manufacturing techniques or **3D printing**. Prototyping applications are discussed in more detail in Unit R040.

Figure 1.18 A physical prototype produced on an additive-manufacturing machine

Where prototypes need to have the functional properties of the finished production component, designers may use CNC machining or additive manufacturing to produce short-run mould tools that can be used to make prototype injection-moulded components.

Error proofing

The optimise phase is also where the designer will carry out **error proofing** on the design. Products or components are often designed to stop them being misused, to prevent them being assembled in the wrong way or to protect the user.

Error proofing as a function of the final product

Perhaps the simplest form of error proofing is to use colour or certain shapes to guide the user about which way round to put things together. For example, a standard 3-pin plug can only be put in a socket one way. This ensures the appropriate connections are made and the user cannot force the plug into the socket incorrectly.

Key terms

3D printing Production of a 3D physical component from a computer-aided design model, by adding material layer by layer.

Error proofing Integration of a mechanism or device into a product or process that stops it being misused, prevents it being assembled in the wrong way or protects the user.

Audio cables are another example; they tend to have different coloured connectors that correspond to the connectors on audio devices. This means that the colours can be matched to ensure correct connections.

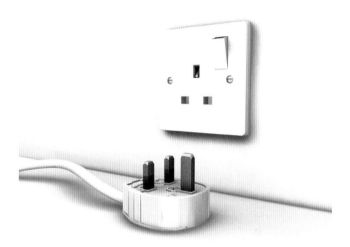

Figure 1.19 A 3-pin plug is an error-proofed product because it can only be inserted in the socket one way

Error proofing during manufacturing

Error proofing is also used in manufacturing to ensure products are assembled correctly. For example, in car manufacturing, there may be colour-coded stickers on one side of a brake disc assembly to show which side should face the front.

In other cases, the geometry of the component is designed so that it can only be assembled one way. It is physically impossible to put the two components together the wrong way because of the position of mounting holes or clips.

In all these cases, this means that errors in manufacturing are minimised, which improves quality.

Error proofing to protect users

Error proofing can also be used to ensure the user cannot harm themselves. For example, a food processor will only turn on when the lid is in place and closed fully. This means the user cannot put their hand in the food processor when the blade is spinning.

Designers also often include **fail-safe mechanisms**, which are features integrated into a product that protect the user from harm in the event of a fault or misuse.

Validate phase

The validate phase is the last phase of the design cycle, where the designer:

- reviews and justifies their design decisions
- tests the design (both market testing and product testing)
- suggests any final modifications to the design prior to it going into production
- evaluates the design solution against the design brief and design specification, to ensure it meets all the requirements set out by the client
- evaluates the impact of the design solution – for example, the social impact, moral issues and the impact of the product on the environment.

Justification of design decisions

Throughout the process of developing a new product, a designer has to take many decisions to ensure the product is developed in line with the specification and meets the requirements of the client.

During the validate phase, the designer may review these decisions and discuss them with the client. In almost all cases, compromises will have to be made to the design. These may be changes to geometry to allow for ease of manufacture, or they could be about managing the balance between cost and functionality.

The designer will evaluate the final design to ensure they have the best possible solution with

the resources that were available and at the most appropriate cost.

Testing the design

Testing a design before full-scale production allows a designer to check whether it functions as intended. Testing can take many different forms, from gaining feedback from potential users or customers to carrying out testing (physical or virtual) of prototypes.

Changes at this stage tend to focus on improvements to the final design rather than major design modifications to multiple variations of a product. However, an exception would be if a designer is working on a critical structural component for a product. Multiple developed designs may need to progress to the validate phase to allow for thorough product testing of different variants, so that the designer can choose the best design. This is a good example of flexibility across stages of the design cycle.

Market testing

A designer undertakes extensive research in the identify phase of the design cycle to gain an understanding of what the market wants from a product. However, it is also helpful for the market to see what the design solution looks like or how it functions.

During the validate phase of the design cycle, the designer or company may share prototypes or **final renders** with the market to gain feedback on the design. This may be in focus groups, where people are invited to test the product and provide feedback. This gives the designer the opportunity to make any minor changes before production.

> ## Key terms
>
> **Fail-safe mechanisms** Design features integrated into a product to protect the user from harm in the event of a fault or misuse.
>
> **Final renders** Realistic images of computer-generated models to show what a finished product looks like without the need to produce a physical model.

Companies can also use the information from market testing to gauge interest in the new product and find out more about the size of the target market or the specific target group the product appeals to. They can compare this to the market research undertaken earlier in the design cycle and ensure the product meets the needs of the users set out in the design brief.

Product testing

There are two types of product testing that can be carried out by designers:

- **Physical testing** is where an actual prototype model is made and tested in the real world.
- **Virtual testing** is where computer-generated models of components, products or systems are created and then simulations are carried out that represent real-world situations.

Physical testing

In some cases, physical testing focuses on improving the function or performance of the product. Multiple variations are made and tested by a range of users, so that the function can be optimised or ergonomic changes made to improve usability.

Sometimes testing needs to imitate real-world situations. Physical tests are conducted that subject the prototype to extreme operating conditions, in order to push a design to its limits and check it can withstand the forces and stresses that will be applied to it during use.

This type of testing often results in the prototype being completed destroyed. For example, in the automotive sector, when a new design of car has been produced, multiple prototypes will be created and destroyed during simulated crashes and accidents. These prototypes are extremely expensive to create, but these tests are required in order to establish the safety of the design so it can be put on sale. Designers need to know that components and products can still function in extreme situations, due to the possibility that lives could be put at risk should they fail.

Designers may use virtual testing to reduce the number of expensive physical prototypes required. In other cases, **non-destructive testing (NDT)** can be used, which checks the integrity of elements of the component – for example, checking for defects in welds on a component.

Figure 1.20 Physical testing – non-destructive penetrant test on a weld

Virtual testing

The ability to run accurate, virtual tests has increased significantly in recent years, with the development of advanced CAD and simulation software packages. In most cases, virtual prototypes do not completely replace the need to produce physical prototypes, but they can dramatically reduce development time and cost.

Key terms

Physical testing Testing undertaken on an actual, physical prototype to see how it reacts to real-world conditions or forces.

Virtual testing Testing undertaken on computer-generated representations of components or products to see how they react to real-world conditions or forces.

Non-destructive testing (NDT) Where components are tested without the need to damage or destroy them.

Virtual tests may be conducted to show how mechanisms work, where assemblies of components are animated to show they function. They may also be used to simulate manufacturing processes. For example, a simulation may be carried out to show how liquid metal flows within a casting process or how plastic flows in an injection-mould tool. This can help designers ensure that the component will be produced without errors.

Research

Research different types of CAD and simulation software and see how virtual prototypes can be checked or tested using simulation.

For instance, two common processes, **finite element analysis (FEA)** and **computational fluid dynamics (CFD)**, simulate how a component may react when subjected to the operating conditions of its environment (for FEA, applying stress, and for CFD, having liquids or gases flow through or around it).

Try to find some specific examples where virtual prototypes have been used in the development of new products. What impact did their use have?

Comparing the design solution against the requirements of the design brief and specification

The development of a new product is a complex process. Designers have to make hundreds of decisions to ensure they have a design that can be manufactured and launched on the market within an appropriate timescale, that functions correctly and that is cost-effective. Therefore, it is possible for a design to move away from the initial brief and specification as decisions are made based on manufacturing considerations or cost.

Designers will take the time to compare and evaluate their final solution against the original design brief and design specification, to ensure

it is in line with the requirements set out in each document.

If there are elements that have moved away from the original specification, designers may have to make last-minute changes or justify why this is the case (for example, due to excessive costs or limitations of manufacturing, changes had to be made to the original specification).

Evaluating the impact of the design solution

Designers have to be conscious of the impact that a new product will have during its manufacture, during its use and at the end of its life. There are many considerations that designers have to take into account, from the social impact of a design to moral and ethical issues associated with the introduction of new products. For example, it may be possible to manufacture a design in a different country at a lower cost, but consideration must be given to the working conditions of employees there.

The impact of the product on the environment is also important:

- Where are **raw materials** sourced, and are they sustainable?
- Does the product produce harmful emissions?
- Can the product be recycled, or does it need to go to landfill at the end of its life?

Key terms

Raw materials Extracted materials that will be processed and then used to produce components.

Finite element analysis (FEA) Method of simulation undertaken in a software package that analyses how a component is affected by applied forces or stresses.

Computational fluid dynamics (CFD) Method of simulation undertaken in a software package that analyses how a gas or liquid flows through or around components and products.

Designers usually undertake a product **life cycle analysis (LCA)** to evaluate the impact of a product on the environment at all stages of its life, from its creation to the point of **disposal**. They have to review the design to evaluate its impact in all these areas and ensure that any compromises that have to be made due to costs or manufacturing can keep any negative impacts to a minimum.

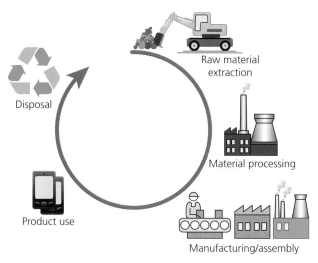

Figure 1.21 Stages of the product life cycle

Key terms

Life cycle analysis (LCA) Technique used to evaluate the impact of a product on the environment at all stages of its life, from its creation to the point it is disposed of.

Disposal Stage in a product's life cycle when it is no longer useful and must be thrown away or recycled.

Test your knowledge

1 What are the four phases of the iterative design cycle?
2 What steps can designers take to understand their target market before developing a new product?
3 Why is it important that designers review their designs against the design brief and design specification during the iterative design cycle?
4 Describe the purpose of making a prototype, and the advantages and disadvantages of virtual and physical prototyping.

Practice questions

1 Give one example of sustainable design. [1]
2 State what is meant by the term ergonomic design. [1]
3 State two methods of primary research. [2]
4 Name two of the categories included in ACCESS FM. [2]
5 Explain one reason why a designer would make a physical model of a design idea. [4]

Topic area 2 Design requirements

Getting started

Figure 1.22 shows a cordless electric kettle. Think about all the criteria that a designer would need to know in a design specification before generating the design of the kettle, including user needs, product requirements, manufacturing considerations, production costs and regulations and standards.

Figure 1.22 A cordless electric kettle

2.1 Types of criteria included in an engineering design specification

Needs and wants

The development of every product starts with the identification of a need. This is typically a problem that needs solving. Design **needs** are communicated through specification criteria created following focused research. These are usually presented as a list of short statements that identify specific needs and **wants**.

- Needs are aspects of the design that are considered to be critical to the future outcome, such as what it must do or be, how it will be used, the level of performance provided by it, and so on.

- Wants are the aspects that are desirable or additional to the main need(s), such as size, shape, colour, and so on.

Needs and wants can be identified by exploring the situation and context, understanding the needs of customers (potential clients, brands, and so on) and knowing the strengths and weaknesses of the competition (existing products).

Within engineering, different terms may be used to explain the relative importance of different specification criteria, such as 'critical', 'desirable' and 'non-essential', or 'must', 'should' and 'could', and designers must fully understand the order of importance of design needs to make sure the product they develop meets these requirements. Critical ('must') criteria will be of a higher priority than other design criteria when final design decisions are made.

Key terms

Needs Aspects of a design that are considered to be critical to the future outcome.

Wants Desirable aspects of a design that are not considered to be critical to the future outcome.

Situation and context

To understand needs and wants best, the engineer must return to the situation and context that has led to the brief:

- The situation is the location where the need for a new design has arisen and where the product will be used.

- The context is the circumstances, such as events that have occurred, leading to the requirement of a design solution.

For example, bathrooms are traditionally designed for people with no mobility issues (situation). Elderly people or people with disabilities can struggle to use them, as getting in and out of the shower or bath can prove difficult with limited mobility (context). Therefore, designers have created a series of mobility aids, such as easy-access baths and showers or supportive handrails, to allow people to get in and out more easily.

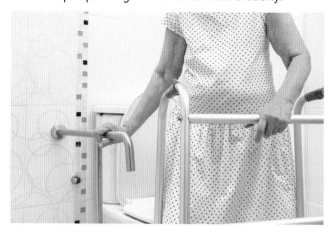

Figure 1.23 Bathroom mobility aids

Needs of the client

It is important to define the exact needs of the client before starting to design a new product. If client needs are not understood in detail, it is easy for work to be based on assumptions.

It is also important to understand who the client is. In some cases, the client may not be the end user, so the needs of the client and of the end user may be very different. This can lead to conflicting design decisions and possible issues later in the design process, which can cause delays and costly design changes to align the design with the needs of the user.

Activity

Outline design specification criteria for a cordless vacuum cleaner. Identify needs and wants using hierarchical criteria statements (such as critical, desirable, non-essential, or must, should, could).

Quantitative and qualitative criteria

From their research, designers will be able to develop design specification criteria that reflect every aspect of the future product. The criteria will have both quantitative and qualitative statements:

- Quantitative criteria are factual and measurable; they are presented as numerical

data in the form of tables, graphs or figures. Quantitative data can be useful for criteria related to size, weight, quantity, cost, and so on, and can be generated from surveys, polls or questionnaires.

- Qualitative criteria are non-numerical, non-factual and based on opinions. They are often presented as descriptions, judgements or preferences that have been gained from trials, questionnaires, interviews, focus group studies or observations.

Quantitative factual criteria tend to be specific and so can be reflected through critical need criteria statements, whereas qualitative opinion-based criteria may be more aligned to desirable want criteria statements.

Activity

Produce quantitative and qualitative criteria statements for the following products:

- electric scooter
- laptop bag
- TV controller
- wall clock.

Product criteria included in the design specification (ACCESS FM)

The design specification will detail the various requirements the product must meet once it has been designed and manufactured. These cover a range of criteria, such as:

- how the product will work
- what features the product will have
- how well a product can carry out its task
- who the target group and intended users are
- where the product will be used
- limitations and constraints of the product
- how the product will look
- how users will interact with the product
- the life cycle of the product
- maintenance of the product
- product safety
- how sustainable the product is.

Important issues such as ergonomics, anthropometrics and **legislation** can be addressed as the specification is developed; manufacturing will follow under its own heading.

ACCESS FM can be used to categorise the criteria included in a design specification.

Key term
Legislation Laws proposed by the government and made official by Acts of Parliament.

A: Aesthetics

Aesthetics refers to how a product appeals to the senses (what it looks, sounds, feels like, and so on). Most products do not have to appeal to every sense. When designing products, aesthetics often refers to a product's shape, form, colour, texture, symmetry and proportion.

For consumer products, designers often aim to produce aesthetically pleasing products because this can be one of the main factors that attracts customers, increasing the popularity of the product.

Some of the most iconic designs achieved their status because they are recognised as being aesthetically beautiful.

Figure 1.25 The Barcelona Chair by Mies van der Rohe is an aesthetically iconic chair

However, a designer should not sacrifice the functionality of the product for the sake of a beautiful design. This means that there needs to be a balance between the function and the form of a product. In engineering design, where the primary focus is to ensure the component or product functions effectively, aesthetics may not be as important. However, many of these functional designs can often be regarded as beautiful.

Designers should define any aesthetic characteristics that they know the user, client or market will desire in the final product and ensure these are set out in the design specification.

Figure 1.24 Apple is renowned for developing aesthetically pleasing products

Figure 1.26 A complex watch mechanism is designed for function yet contributes to the aesthetics of the watch

Many brands have a design language that ensures their new products are familiar to their customers. This could include colour, material, shape/geometry, **surface finish** (for example, painted or polished) and any branding/logos.

Apple has built a reputation for developing products where colour and branding are central to its designs. This has contributed to their appeal to customers. Apple changed the appearance of home computers in 1998 with the introduction of the iMac, which was available in a range of bright colours. It has continued to push this trend in recent years by developing a range of smart phones and tablets in multiple colours.

Apple's design language has changed over time and, alongside its use of colour, it has also defined the appearance of its products through the materials they are made of and the associated manufacturing processes. For example, the **unibody** design of the MacBook is machined from solid aluminium, which demonstrates the connection between aesthetics, function and manufacturing that can be achieved through integrated design.

Figure 1.27 The 1998 Apple iMac

Key terms

Surface finish Nature of a surface, defined in terms of its roughness, lay (surface pattern) and waviness (irregularities in the surface).

Unibody The frame is integrated into the body construction; every panel provides part of the structural design.

C: Cost

The two biggest constraints on the development of new products are timescale and budget. There may be a limit on:

- the time available to bring the product to market before the competition
- the amount of money available to bring the product to market.

A development budget is how much money a company is willing to spend on a new product. It includes all the costs required to complete the design and ensure it is ready for full-scale manufacture and sale.

Development budgets can vary hugely across different products and industries, but they always need to be defined and monitored carefully.

A range of costs needs to be considered. These should be defined and calculated accurately in discussion with the client. These costs include:

- market research
- staffing
- prototyping
- testing
- manufacturing setup.

Development costs will have a direct impact on the selling cost of the product. The designer will need to discuss with the client what the **target cost** of the product is – that is, the price at which the client wants to sell the product to the customer.

The designer and client also need to discuss the number of products that are expected to be sold. This is so the business can calculate how many products they need to sell to cover all the development costs and begin to make a profit. The point where enough products have been sold to earn back the development costs is called the **break-even point**.

If the business is not going to reach break-even point, it may have to review the selling cost. A higher cost would increase the price to the consumer and might make the product too expensive for the market. If these costs are not calculated, a product could be launched onto the market that would not generate enough sales, and therefore make a loss for the business.

It is essential that costs are controlled to maintain the selling price (which may be very important for the success of the product).

Key terms

Target cost How much a company wants to sell a product for when it is put on sale.

Break-even point The point where enough products have been sold to cover the development costs.

Demographic Used to describe the numbers and characteristics of people who form a particular group.

C: Customer

The design specification sets out the intended users and customers for the product, or the target market – that is, those people who will be most likely to buy and use the product. Research carried out in the design cycle will define the characteristics of these people. The criteria may include:

- age range
- gender (where relevant)

- lifestyle information – for example, leisure habits and income
- geography (where the user lives or the location of customers)
- buying habits – for example, brand loyalty, when a consumer will continually buy products from a particular company over a long period of time due to their commitment to that brand.

Most products are designed and developed to meet the needs of a specific group, or target audience. The target audience may also be the **demographic** that created or identified the gap in the market that resulted in the need for the product. Therefore, a business needs to understand exactly who their target audience is and the characteristics of that audience. In some cases, the target audience will include lots of different groups of people, and in other cases the target audience may focus on specific individuals or people in particular situations.

By developing a detailed understanding of the target audience, the designer can ensure that the product not only appeals to that particular group aesthetically but also that it meets their needs ergonomically and can function in the way they expect.

Importantly, designers have to be aware of inclusivity and should not aim to develop a product that might be offensive or exclude a particular group of people. Even for products specifically targeted at a key group – for example, based on gender – designers have to be aware of how the design will be received by different people within each category and be as inclusive as possible.

Activity

Try to define a profile of the target audience for each of the following products:

- wireless baby monitor
- games console
- cordless vacuum cleaner.

You may want to consider the list of criteria provided earlier to help you.

E: Environment

The working environment is where the product will normally be used. However, it is also important to prepare for situations where the product may be exposed to conditions outside of this. Cars are designed to ensure they can continue to function in hot or cold weather, even when the temperature changes are dramatic.

Defining the working environment in the design specification is critical to the success of the design. It can have an impact on the material that is used, the components included in the design and the way the product is assembled. For example, if the product needs to function underwater (as with underwater cameras or watches), then any outer casing must be fully sealed to prevent water getting into the product.

When designing products, designers also think about sustainability and the long-term effects of their products on the environment. A focus on sustainability should aim to ensure that meeting the needs of the present does not prevent future generations meeting theirs. This can apply to economic growth, the social make-up of societies and **supply chains**, as well as the use of energy and material.

Designers need to consider sustainability at all stages of a product's development. It can help to assess all the stages of the life cycle of a product. The designer will need to think about the future effects of the product, in particular the materials and the energy involved in the product's manufacture, use and disposal. An example could be using recycled materials and renewable energy in the development and manufacture of the product.

Designers usually undertake a product life cycle analysis (LCA) to evaluate the impact of a product on the environment at all stages of its life, from its creation to the point of disposal (see page 23).

In a design specification, criteria will be set for the development of the product in relation to sustainability. These may include:

- reducing the amount of materials used in the product

- using specific types of material – for example, non-toxic/recycled/recyclable
- increasing the reliability of the product to minimise replacement and disposal
- designing for maintenance
- reducing energy use or emissions during manufacture.

Key term

Supply chain Network of businesses that supply materials, components or services needed for the manufacture of a product.

Stretch activity

Write a series of specification points for a sustainable version of a product of your choice.

S: Safety

Product safety requirements should be considered throughout the design specification, with detailed information on features that must be incorporated into the product to protect the user. This information should cover the way the user will operate the product, the user's needs and where the product will be used.

Depending on the product, various factors may affect safety. The designer and manufacturer must also follow certain regulations before the product is sold. Products should only be put on sale if they comply with safety regulations.

For example, the General Product Safety Regulations 2005 (GPSR) require all products to be safe during normal or expected usage. If a product does not comply, authorities have the power to take appropriate action. Manufacturers must ensure they:

- minimise any risks associated with the product
- keep records of technical documentation
- label the product appropriately
- give instructions on how to use the product safely.

There are also specific regulations for particular business sectors, covering a wide range of products and industries. This includes electrical and electronics, machinery, furniture and toys. Information on the regulations can be found on the UK government website, 'Product safety advice for businesses' (**www.gov.uk/guidance/product-safety-advice-for-businesses**).

For industries such as aerospace, where there is the potential to cause significant harm or death to large numbers of people, there are specific authorities and many **standards** for ensuring the safety of users. Extensive testing, research and development are carried out by companies in these sectors to ensure that they are compliant with standards and that customers are safe.

A particular focus of legislation is to ensure products are safe to use. Designers and manufacturers have to make products that comply with relevant standards defined in legislation. If a company, designer or manufacturer produces a product that subsequently breaks the law, they can be **prosecuted**. Legislation and standards are covered in more detail later in this unit, in section 2.3.

Companies regularly highlight how their products have been designed to protect consumer safety for the benefit of their reputation and sales. Table 1.1 shows methods a company may use to demonstrate the safety of their products.

Key terms

Standard An agreed way of doing something, such as making a product, managing a process or delivering a service.

Prosecuted Officially accused in court of breaking a law.

Activity

Look at Figures 1.28 and 1.29. How have these products been designed to maximise safety?

Figure 1.28 A food processor

Figure 1.29 An electric kettle

Table 1.1 Examples of safety-specific product requirements

Method	Description	Example
Compliance to standards or regulations	Stating that a product meets safety standards or regulations gives customers faith in a product.	Cars are crash-tested to check they conform to standards and are safe before sale.
Fail-safes	If a product develops a fault, it automatically enters a safe state to protect the user.	Brakes on a lift operate automatically if power is lost.
Specific safety features	Specific features are integrated into a product to improve safety.	Car seatbelts are added to cars as a specific feature to improve safety.
Product reliability	Ensuring a product is reliable over time aims to prevent failure.	Aircraft engines are tested by having objects fired into the engine to ensure they can continue to work in the event of a bird strike.
Warnings, labels and instructions	Warning labels or clear instructions ensure the user understands the dangers and how to use the product correctly.	Warning stickers are placed on exercise equipment where the user could trap parts of their body.
Error proofing	Products are designed so that they cannot be used incorrectly.	A 3-pin UK electrical plug can only be inserted into the socket one way.
Hygiene	Components can be cleaned easily or made single use.	This is essential for many medical products, such as gloves.
Child safety	Avoiding the use of small parts and harmful materials or coatings can help to protect children.	Toy bricks for toddlers are made larger to avoid choking.
Ease of use	A product that is easy to operate is safer than one that is difficult to use.	Automated controls on a plane allow the pilot to focus on critical elements in the cockpit.
Safe modes	A product goes into a mode where it can still be operated but with limited functionality.	In some vehicles, a safe mode is activated if there are errors with the engine, allowing the driver to move to a safe place.
Material selection	Materials are selected that are non-toxic, can resist corrosion, are not flammable and can cope with stresses.	Non-flammable material is used in furniture design.

S: Size

There is normally a series of limitations and constraints that the final design cannot exceed. These can restrict the freedom the designer has over the final design.

The product may need to have particular performance criteria, as discussed earlier, but a constraint on the product may limit the flexibility of the solution. Therefore, a compromise will need to be made to maximise performance within acceptable constraints. For example, a component may need to be a particular size so that it can fit within the assembly of the product but there may also be a limit on how heavy it can be.

The limitations and constraints can have an impact on the materials used and the associated manufacturing processes, which then affects the shape of components. Size, weight and functional limitations are common constraints – products cannot be much too big or much too small; otherwise they will not sell. Functional limitations set the maximum and minimum performance criteria for the product. For example, if a product needs to be waterproof, the maximum depth may be defined. When a range of different limitations and constraints apply to a product, a range of solutions will need to be produced and tested until a design is found that works best.

Tolerance

When a component is manufactured, it will typically vary in size. In some cases, this variation is extremely small but this still needs to be taken into account. A **tolerance** is the amount of variation allowed in a dimension. If the dimension is within tolerance, it is accurate enough to ensure the product can fit together and function properly. The more accurate the component needs to be, the more difficult it is to make. Generally, this then makes the component more expensive to produce.

Different manufacturing processes may produce components with a variety of tolerances. If a metal bar is cut by hand with a saw, the final result may not be as accurate as one machined on a lathe.

Tolerances must be included in the design specification, as well as on engineering drawings of the components.

See Communicating design outcomes (page 78) for more examples of tolerances.

Key term

Tolerance Amount of variation allowed in a given dimension.

Stretch activity

In a workshop, take some material and cut it to a length of 30 mm. You may use any material that you have available – for example, you may use cardboard and cut it by hand using scissors or a craft knife, or some wooden dowel and cut it with a saw.

Once you have done this, measure the length of the piece you have cut as accurately as possible. How close to 30 mm was the piece you cut? Based on this, what would be a suitable tolerance for the measurement using the production process you chose?

Ergonomics

When designing for people, designers must ensure that the dimensions of their products are appropriate for the people that will use them. Ergonomics is the study of how people interact with the objects they use and the environments they use them in. It is partly to do with the shape, size and geometry of a product, but it also considers how easy a product is to understand and use. An ergonomically designed product should be comfortable to use, easy to understand and fit the user it is designed for.

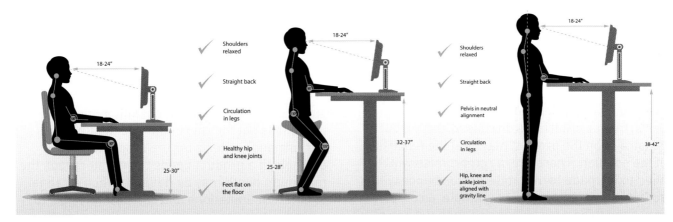

Figure 1.30 Ergonomic considerations associated with the use of a height-adjustable desk

Humans vary in size dramatically, so many products are adjustable, so that they can be adapted to a range of different users. Figure 1.30 illustrates some of the considerations that need to be taken into account to ensure the design of a height-adjustable desk promotes good posture from the user. If the desk is designed ergonomically, the posture of the user will not cause any strain or associated injuries.

The design of some products appears to be extremely simple but they are perfectly designed to be ergonomic. For example, consider a toothbrush or cutlery. These have been designed to fit in your hands, and to allow you to grip them effectively and to easily manipulate them into a range of positions.

Figure 1.31 A toothbrush is designed to be comfortable to hold and easy to manipulate into a range of positions

Some products are more complicated – for example, in a car there are lots of controls and the interface between the user and the vehicle becomes much more complicated. However, when the car is well designed, the position of the controls and the driver means that all controls are easily accessible. The user can remain seated in comfort for long periods of time with their main focus on driving the car safely.

Figure 1.32 A car interior is ergonomically designed to ensure all the controls are easily accessible from a comfortable driving position

Activity

Consider the design of an office chair. Discuss the different ways it can be adjusted and how this contributes to the ergonomics of the design.

Anthropometrics

Anthropometrics is the study of the measurements of the human body. Anthropometric measurements are taken from people of different ages, genders and sizes and collated into tables and diagrams, so that designers can ensure products are the correct size for the user they are being designed for.

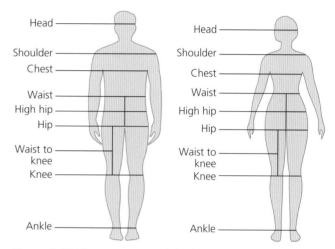

Figure 1.33 Measurements of the human body used for anthropometrics

Anthropometric data includes measurements for dimensions such as height, finger length, hand size, reach and grip. These measurements are grouped into percentiles, which include the 5th percentile (the smallest 5 per cent of people are below the 5th percentile), the 50th percentile (the average size of people) and the 95th percentile (the largest 5 per cent of people are above the 95th percentile).

Designers need to consider the range of users that the product is being designed for and ensure that they develop the product to be suitable for this range. For example, seats on public transport may be used by people across all percentile ranges, so designers have to make the seat design as suitable as possible for all users.

Anthropometric measurements of users will be detailed in the design specification so that the designer can reference them when designing the product.

Table 1.2 Anthropometric data for standing (19–65 year olds)

Dimensions (mm)	Men (percentiles)			Women (percentiles)		
	5%	50%	95%	5%	50%	95%
Height	1,630	1,745	1,860	1,510	1,620	1,730
Eye level	1,520	1,640	1,760	1,410	1,515	1,620
Shoulder height	1,340	1,445	1,550	1,240	1,330	1,420
Elbow height	1,020	1,100	1,180	950	1,020	1090
Hip height	850	935	1,020	750	820	890
Knuckle height (fist grip height)	700	765	830	670	720	770
Fingertip height	600	675	730	560	620	680
Vertical reach (standing position)	1,950	2,100	2,250	1,810	1,940	2,070
Forward grip reach (standing)	720	790	860	660	725	790

Source: www.ocr.org.uk/Images/304609-specification-accredited-a-level-gce-design-and-technology-h404-h406.pdf

Activity

Explain why anthropometric data is compiled from measurements of different groups of people and how it can be used by designers.

Stretch activity

Using the data in Table 1.2, consider the checkout in a supermarket.

Which dimensions would be essential to ensure any customer at the supermarket can easily pack and pay for their shopping?

Figure 1.34 A supermarket checkout

F: Function

The function of a product is what it will do and how it will work. This will be detailed in a design specification as a requirement of the finished product.

Customers may be attracted to products because of their appearance, but the form of a product should not negatively affect its function. Customers usually buy a new product because it helps them to carry out a task more efficiently and makes a difference to their lives. Designers must ensure that the product carries out its function as effectively, reliably, simply and safely as possible.

In industrial design, and particularly in engineering design, the functionality of a product can be seen as more critical to successful design than creating an aesthetically pleasing product. However, the best designs tend to find the perfect balance, with products performing their function in the simplest, most reliable and most efficient way possible, as well as being pleasing to look at.

Activity

The products in Figures 1.35 and 1.36 have the same function: to squeeze juice from lemons. Has one been designed with a greater focus on function or form? Explain why you think this.

Figure 1.35 Philippe Starck Juicy Salif lemon squeezer

Figure 1.36 Traditional lemon squeezer

The features of a product are the distinguishing characteristics that make it appealing to customers. The client or user may have asked for specific features in the design of a new product during the development of the brief or because of market research. Sometimes, the features of a new product may be introduced in response to competitors' products that have new or more advanced features in their most recent versions.

The continued technological developments in smart phones are a good example of how new product features are continually added to a **multi-functional product** to make it more competitive against rival products. Smart phones have increasingly advanced cameras, provide the ability to make contactless payments, monitor the user's physical activity and act as portable entertainment units.

The design specification must include all the features required in the product so that they are incorporated into the final design.

Design specification criteria will include a set of **product performance** criteria that indicate how

well a product must carry out its task. These criteria can refer to a range of different areas: they may cover performance during use or the product's ability to withstand conditions in its working environment.

In order to define the performance criteria, it is necessary to understand how the product operates and where and how it will be used. The performance criteria can affect which materials or components are selected to be included in a product. They can also influence the shape and dimension of individual components because these can positively or negatively affect the product's strength or **durability**.

Key terms

Multi-functional product Single product that can carry out the tasks of multiple products.

Product performance How well a product can carry out its task.

Durability Ability of a material to withstand wear, pressure or damage.

The following are examples of performance criteria for a component or product:

- weight
- strength (**tensile strength**, **compressive strength**, impact strength)
- strength-to-weight ratio
- **resistance to corrosion**
- water resistance
- operating temperature
- number of times it can operate before failure
- durability
- flammability.

An example of how performance criteria have an impact on the development of a product can be seen in industrial piping applications. In some cases, **composite materials** are used instead of metals to produce piping. This reduces the weight of the piping, which makes transportation easier, and increases resistance to corrosion, which extends the life of the component.

M: Materials

When developing a product, a designer will identify materials that are fit for their intended purpose. They must be able to perform reliably and be easily available at an acceptable cost. For example:

- **Low carbon steel** is an ideal material for making car body panels because it has outstanding **ductility** and **toughness**, and very high strength-to-weight ratio. It is **malleable**, machinable, weldable and relatively inexpensive to produce; although it can be subject to corrosion and rust, it can be easily recycled.
- Pure silver is considered the best conductor of electricity but it is prone to tarnish, which can affect the distribution of current; it is also very expensive, so it is not widely used in electrical applications.

Figure 1.37 Composite materials used to make piping for industrial applications

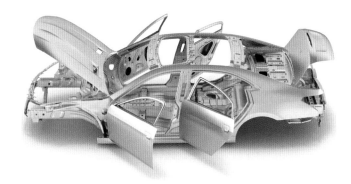

Figure 1.38 Low carbon steel body panels

Key terms

Tensile strength Strength of a material when it is stretched or pulled.

Compressive strength Strength of a material under load (when the load is 'compressing' the object).

Resistance to corrosion Ability of a material to resist deterioration caused by reactions to its surrounding environment.

Composite materials Materials made up of two or more different materials, combining their properties to create a new, improved product.

Low carbon steel A low carbon ferrous material (contains iron) that consists of less than 0.3 per cent carbon; also known as mild steel.

Ductility The ability of a material to be stretched under load without breaking.

Toughness The ability of a material to resist impact or shock loads (such as press-forming a car body panel).

Malleability The ability of a material to be shaped or deformed by compressive forces (such as hammering or pressing).

When selecting materials for engineered products, conflicting issues will need to be assessed, such as:

- aesthetic impact and how the material appeals to the senses (how it looks, sounds, feels, and so on)
- working properties, which will indicate how materials behave in use:
  - mechanical properties (such as hardness, toughness, elasticity, malleability, ductility, tensile strength), which describe how a material behaves when subjected to different types of loads
  - electrical properties (such as electrical conductivity, electrical resistivity), which describe how a material behaves when subjected to electrical potential difference
  - thermal properties (such as thermal conductivity, thermal expansivity), which describe how a material behaves when subjected to temperature changes
- cost, which can be dependent upon a range of factors, including:
  - the scarcity of the raw materials
  - how much processing is required to turn the raw materials into a usable material in a usable form
  - the form in which the processed material is available for supply (such as granules, ingots, bar, sheet and plate, pipe and tube); many materials are available in **standard stock sizes** and quantities (see 'Material availability and form', page 50)
  - the quantity of material required – buying materials in larger quantities can reduce the cost (due to reduced processing, handling and transportation), but this will need to be balanced against the cost of storage prior to use
- ease of manufacture, which will point to how the product could be manufactured (such as cast, formed, machined, fabricated) (see page 41)
- environmental impact and considerations for the product's end of life – balancing the needs

of sustainable use of resources and what happens to products when they come to the end of their life is a challenge that needs to be addressed so that there is minimal lasting effect upon the environment.

Activity

Research a range of engineering materials and their availability in standard stock sizes.

Eco-materials

Designers are increasingly trying to use eco-friendly materials in order to minimise the number of products that are not biodegradable or that are sent to landfill at the end of their life.

Particular attention has been given to reducing the amount of plastic that is sent to landfill, as it takes hundreds of years to degrade.

Bioplastics, such as corn starch polymers, have been created for products such as drink containers or single-use cutlery. Starch-based polymers can be extracted from vegetables that are high in starch, such as potatoes. The polymers decompose in a short period of time and are made from natural sources so do not harm the environment. Plant starches have also been successfully used to make fabrics that are fully biodegradable, to replace synthetic fibres such as polyester. In addition, plant fibres such as hemp have been successfully used in composite materials, alongside natural plant-based resins from biological sources, such as soy bean or corn.

Key terms

Standard stock sizes Materials and components that are readily available in a range of sizes (such as sheet material at 60 cm x 60 cm, 1.22 m x 2.44 m, and so on; nuts and bolts at the following standard sizes: 4 mm, 5 mm, 6 mm, 8 mm, 10 mm, 12 mm, and so on).

Bioplastics Biodegradable plastic materials produced from renewable sources such as corn starch, vegetable fats and oils.

Recycled materials

A wide range of materials can be recycled when the product they were originally used for reaches the end of its life. Many products are now manufactured using recycled materials, which reduces the need to extract raw materials from the ground. Many types of metal, polymer, glass, textile and paper can be sourced in recycled form.

Figure 1.39 Scrap metal for recycling

Figure 1.40 Used plastics collected for recycling

Recycled metal

A large amount of metal used in industry comes from recycled sources. Many ores containing metal are becoming increasingly scarce, so using metal that is recycled is more cost-effective and also reduces the need to cause further environmental damage through mining.

Most metals can be recycled again and again, and although the process of recycling metal uses lots of energy, this produces far less carbon dioxide than mining raw material from the ground and processing it. Metal is collected by specialist companies and separated into its various forms. The metal is then melted at high temperatures and cast into ingots that can be used for new applications.

Recycled polymers

A large number of polymers can also be recycled. Polymers are manufactured from crude oil, which is becoming increasingly rare. They also do not biodegrade, so when they are disposed of, they can cause problems for the environment. Using recycled plastic avoids unnecessary disposal.

Plastic products tend to include symbols that identify the type of polymer they are made from. It is important to know from which type of polymer a product is made because not all types of plastic can be recycled.

Plastic can be broken down into small chips that can then be reused to make new plastic products using processes such as injection moulding.

Recycled glass, textiles and paper

Glass, textiles and paper can also be recycled:

- Glass can be continually recycled; it is crushed and melted before being moulded into new products.
- Many paper-based products are now manufactured from recycled material.
- Many fibres and fabrics can be easily recycled, which is increasingly important as the textile industry, particularly the fashion sector, continues to grow.

New and emerging materials

Advances in technology, science and manufacturing processes can result in the development of new materials with different properties that can help to solve design problems or improve the performance of products. Designers continually explore how these materials can be incorporated into products to take advantage of these improved properties.

Modern materials

Many materials defined as 'modern' may have been created many years ago through scientific exploration. This includes materials created for space exploration, due to the harsh conditions faced by equipment and astronauts when they leave the Earth's atmosphere. Designers may then find applications for the material in more everyday items.

For example, titanium is a metal that has been used in space rockets and Formula 1 cars for many years. Titanium is extremely strong but lightweight and is therefore used where strength and weight are essential design factors. As well as being used in the aerospace and automotive industries, titanium has also been used in consumer products, such as smart watches, to make strong, durable but lightweight casings.

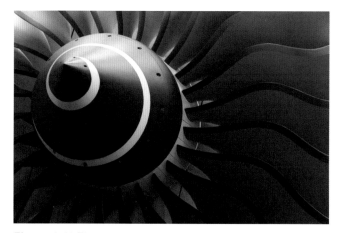

Figure 1.41 Titanium is used in jet engines in the aerospace industry

Graphene is a material manufactured from a single layer of carbon. It is stronger than steel, can conduct electricity, is transparent but is also extremely lightweight. It can be used as coatings for products or in the development of flexible display screens due to its lightweight, flexible and conductive properties.

Nanomaterials are also being used as coatings for products. Nanoparticles are extremely small, which means that when they are added to a product, they do not increase its weight. They can also have antibacterial properties, so they are integrated into medical devices and materials that can be used to help treat diseases.

Smart materials

Smart materials change in response to stimuli in the environment:

- Shape memory alloys (SMAs) remember their original shape and can return to this state when heated if they have been bent out of shape. They can be used in a range of applications, including frames for glasses or as components in machinery.
- Thermochromic materials change colour in response to variations in temperature and can be used in products where awareness of temperature is critical – for example, in drinks containers for babies or young children that change colour to indicate that the contents are too hot. Thermochromic pigments can be included in products by incorporating them into paints, textiles, inks or plastics.
- Photochromic materials change colour in response to variations in light intensity. They can be used in windows that can become darker, lighter or change transparency depending on how much light there is. They can also be used in sunglasses that darken as sunlight increases.

Figure 1.42 A cup made from a thermochromic material

Key term

Smart materials Materials that change in response to stimuli in the environment.

Composite materials

Composite materials are made up of two or more different materials, combining their properties to create a new, improved product. Some composites have been used for a long time:

- Concrete is a combination of cement, sand and stones that when mixed with water hardens to create a material with excellent compressive strength.
- MDF (medium-density fibreboard) is created by combining wood fibres with resin to make wood panels for use in applications such as furniture making.

Developments in composite technology mean their use has increased in recent years. The combining of different materials in composites can provide performance benefits, such as the weight-saving and strength properties of carbon fibre.

Carbon fibre has been used in Formula 1 cars since the early 1980s. However, production has traditionally been time-consuming and expensive. Recent developments in manufacturing processes have resulted in the use of carbon fibre becoming more widespread. Lightweight cars produced from carbon fibre allow for more fuel-efficient vehicles, or for heavy batteries to be incorporated into vehicles with less compromise on performance. Carbon fibre is also used to produce sports equipment and helmets.

Figure 1.43 Carbon fibre components in a high-performance racing car

Improvements in materials

There are many different materials with many different properties. Some properties can be an advantage in certain applications but a disadvantage in others. Designers continually consider how improvements in materials can be incorporated into designs to improve performance.

Huge amounts of time and resource go into the research and development of new materials. For example, the aim may be new materials with the same strength properties but lighter in weight or to improve their ability to cope with high temperatures.

The automotive industry is a good example of this. In the past, cars would have included large amounts of steel in their construction. Steel is extremely strong but heavy, which makes cars less fuel efficient. As automotive manufacturers have tried to reduce emissions and fuel consumption, they have looked for lighter materials. Many cars now use aluminium for the main vehicle chassis, which dramatically reduces weight. As cars move to electric power, the industry will need to continue to reduce vehicle weight due to batteries currently being very heavy. This has seen the introduction into vehicle design of composite materials such as carbon fibre.

Materials for prototype modelling are covered in detail in Unit R040 (page 211).

Figure 1.44 Woven carbon fibres that are used to create lightweight but strong components

Activity

Do you have any creative ideas for items that you could upcycle? Try to create a design using upcycled materials or products.

Case study

Figure 1.45 The BMW i3

The use of carbon-fibre-based materials has traditionally been restricted to low-scale production, due to the process being **labour intensive**, which makes the components expensive.

However, the BMW i3 is one of the first examples of a mass-produced product manufactured using composite, carbon-based materials, in particular carbon fibre reinforced plastic (CFRP). The i3 is a hybrid vehicle; the use of CFRP keeps its body weight to a minimum, which compensates for the heavy batteries required when the car is running on electricity. This makes the car more fuel efficient.

Manufacturing

A design specification will include considerations relating to the manufacture of a product. Some of these considerations may be fully defined before the design has been developed, but others may develop when areas are finalised, such as the shape and size of the design, its function and assembly of the design.

Before developing a design, the designer may know that the product must be made from a particular material or may have decided the scale of production based on analysis of market need. In other cases, the designer may have more freedom to make decisions. Either way, the design specification will provide all the manufacturing information that the designer needs to know before starting development.

Ease of manufacture

During the development of a new product, designers need to continually consider how the product will be manufactured and include elements in the design that make the process of manufacture as easy as possible. This can reduce production costs and improve the speed of assembly. There are many ways that designers can achieve this, from the use of **standard components** to ensuring the design has suitable geometry for the selected manufacturing process.

Standard components

Standard components are designed and produced to specific and 'standard' sizes. Because these dimensions are known, they can be integrated easily into the design of products. Examples of standard components are nuts and bolts, washers, bearings, screws and rivets.

Standard components are mass produced and easily available, which makes them cost-effective. They can also be assembled or disassembled using commonly available tools, making maintenance, repair, assembly during manufacture and disassembly at the end of life much easier.

Pre-manufactured components

In many designs, there will be components and **sub-assemblies** that are manufactured separately and then assembled into the final product. In some cases, these may be produced by subcontractors and arrive at the manufacturer to be integrated into the final product during assembly. This may be because the subcontractor has greater expertise in the production of these items or because the manufacturer does not have the capability to produce them.

Pre-manufactured components and assemblies are used regularly in the automotive industry. Companies with expertise in upholstery may produce pre-manufactured components for car interiors, such as seats, and companies specialising in electronics may produce elements of the dashboard.

Figure 1.46 A car seat may be pre-manufactured by a specialist company

Figure 1.47 An electronic dashboard assembly may be produced as a pre-manufactured component

Key terms

Sub-assemblies Units of assembled components designed to be incorporated with other units and components into a larger manufactured product.

Pre-manufactured components Components or sub-assemblies manufactured separately from the whole product, sometimes by an external supplier, that are then assembled into the final product.

Design for manufacturing and assembly

Design for manufacturing and assembly (DFMA) is a design principle that combines two elements: design for manufacture and design for assembly. This ensures that any component or product that has been designed can be successfully manufactured and then assembled in a way that is cost-effective and simple, while reducing waste processes and ensuring quality.

Design for manufacture considers the material, cost and manufacturing process for a component or product throughout all stages of the design. This ensures that the material selected is fit for purpose, that costs can be managed and that the geometry of the component can be produced using the selected manufacturing process.

Let us consider two common manufacturing processes: injection moulding and CNC machining. Injection moulding is explained on page 57. When designing for these processes, the geometry of components needs to be considered. The following are key design considerations for injection-moulded components:

- Use consistent wall thickness – the walls of an injection-moulded component should have a **uniform thickness**. Thick sections or changes in thickness should be avoided because this can result in **part warping** as these different sections cool down at different speeds.

- Ensure the edges of the component are round – this ensures the material can flow into the mould properly, resulting in a better finish, and rounded corners reduce stress compared with sharp corners.

- Hollow out thick areas where possible and add ribs to thin areas for strength, to ensure the components maintain a consistent wall thickness. This avoids warping and sink marks.

- Add **draft angles** to the walls of components – this allows the part to come out of the mould. Think of how the sides of sandcastle buckets or ice-cube trays are angled so that the sand or ice can easily fall out.

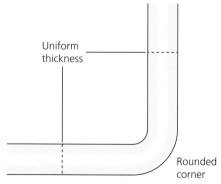

Figure 1.48 An injection-moulded component

Figure 1.49 Consistent wall thickness in an injection-moulded component

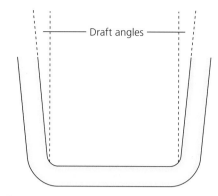

Figure 1.50 Draft angles in an injection-moulded component

Key terms

Uniform thickness Consistent thickness throughout the whole component.

Part warping When a moulded component deforms from its desired shape because parts of the material cool and shrink at different speeds.

Draft angle Sloping face on the wall of a component, set at a specific angle so that it can be removed from a mould.

For CNC-machined components, it is important to consider the following:

- depth of pockets – tools used in machining have a limited cutting depth

- radius of corners – cutting tools come in set diameters

- hole sizes – the size of holes should align with standardised tools

- wall thickness – if a wall becomes too thin, it can be affected by vibration and affect the surface finish.

Figure 1.51 A CNC-machined component

There are also many design for assembly considerations, such as:

- Keep the number of components as low as possible.
- Use clips or locating pins – clips so that components clip together easily without the need for tools; locating pins to ensure the components are accurately placed when assembled.
- Use temporary fasteners, such as screws or nuts and bolts, if a product needs to be disassembled for repair or maintenance.
- Consider the position of any fasteners – they should be positioned so that they can be accessed by tools during assembly. If not, this makes assembly more complex, which adds time and cost to the process.
- Use the same components and tools throughout the assembly process.
- Make it obvious which way parts should be assembled. Components can be designed so they can only be assembled one way or so the correct orientation of the component is obvious. In a USB port, the plug can only be inserted one way. (See 'Error proofing' on page 19)

Design for disassembly

Design for disassembly considers how easily products can be taken apart. This allows:

- critical components to be cleaned if they become dirty with use
- products to be maintained to avoid faults developing

- products to be repaired if a fault occurs (for example, replacing critical components)
- components and materials to be separated at the end of a product's life, so that they can be reused, recycled or disposed of correctly.

The design of components must be considered to ensure disassembly is possible:

- Avoid using permanent fixing methods. Processes such as welding or the use of adhesives to fix components together make disassembly extremely difficult.
- Use temporary fixing methods. Using standard components such as screws or nuts and bolts means the product can be taken apart easily. Common tools can be used to disassemble the product, and the product can be reassembled if it is only being taken apart for maintenance.
- Consider clip-together features. These are common in plastic moulded components – for example, batteries can easily be replaced in a remote control by unclipping the cover. (However, in some cases, these components can be designed so that once clipped together they are very difficult to take apart.)

Figure 1.52 A clip-on battery cover on a remote control allows for easy replacement of batteries

Key term

Design for disassembly Features added to a design that allow it to be easily taken apart for cleaning, maintenance or disposal.

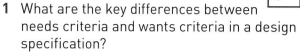

Robotic manufacturing and assembly

Robots have been used in manufacturing alongside humans for many years to assist with a range of processes, such as picking up components and assembling them, or carrying out dangerous processes such as welding.

Robots are a huge advantage in manufacturing as they can carry out, without errors, repetitive processes that can become demotivating for a human. Advances in robotic technology mean that robots can now perform increasingly complex tasks, leading to many factories becoming almost completely automated.

The increased use of robots can be seen in the automotive or consumer electronics industry, where very few humans can be found inside factories:

- Robots are able to assemble vehicle body panels, bond them together and then paint them without human interaction. Employees are present to carry out maintenance and deal with faults or to carry out finishing processes that are still too complex for robots, such as installing the interior or checking for blemishes in the paint.

- In the electronics industry, where a clean environment is essential, pick and place robots can assemble complex electronic products much more accurately and faster than humans, without the possibility of contaminating the sensitive components.

Technologies such as artificial intelligence and increased digitisation mean the use of robotics in manufacturing is likely to continue to grow. Robots are also becoming more affordable, which will make them more accessible to a wider range of businesses. However, the current investment cost is still extremely high, which means they are still mainly used in large-scale mass-manufacturing environments.

Test your knowledge

1 What are the key differences between needs criteria and wants criteria in a design specification?
2 What do you understand by the following terms:
   - situation
   - context?
3 What design specification requirements must a product meet once it has been designed and manufactured?
4 Outline a typical working environment for a wireless charging station.
5 Using one product as an example, show how improvements in materials can benefit consumers.
6 What are the main design-for-assembly considerations?

2.2 How manufacturing considerations affect design

Designers need to consider many different factors when designing new products or re-designing an existing product. Perhaps the factors we most commonly think of as important are the features and functions of the product and its aesthetic appearance. However, there are other key factors concerning how the product will be manufactured and in what quantity.

The relationship between design and manufacturing is a two-way process. The design will influence which production processes and manufacturing methods are most suitable to meet the design requirements of the product. But the reverse is true too: the design will be affected by the suitability and availability of different production processes and manufacturing methods.

Scale of manufacture

Let's start by considering how many of the products will be manufactured each time. Will there be only one, several or a large quantity? This is often called the scale of production.

There are three types of **production method** related directly to scale of production. These are:

- one-off production
- batch production
- mass production.

The most suitable type depends on the scale of production. Of course, products often contain lots of different components, which are then assembled to make the final product, so a range of different production methods might be required for a single product.

One-off production

One-off production is used to manufacture products one at a time. Sometimes these products are called 'bespoke', 'custom' or 'unique', as they are usually made to meet a very specific need of a customer or user. One-off production is also known as 'job' or 'jobbing' production.

Examples of products made using one-off production include towers (Figure 1.53), bridges and stadiums.

Figure 1.53 The Eiffel Tower is an example of one-off production

Another example of products designed for one-off production is specialist transportation, such as high-performance cars and private aircraft. The Bloodhound SSC car (Figure 1.54) was manufactured using one-off production techniques.

> **Key terms**
>
> **Production methods** How things are made; there are three main types: one-off production, batch production and mass production.
>
> **One-off production** Manufacturing products one at a time.
>
> **Bespoke** Made specifically for a particular customer or user.

Figure 1.54 The Bloodhound SSC car is an example of a one-off unique product

Note: A product does not need to be unique to be manufactured using one-off production techniques. One-off production can be used to produce several similar items – just one at a time.

Prototypes can also be made using one-off production.

Although a final overall product might be manufactured by one-off production, many of its component parts might have been made using different techniques, such as batch or mass production.

One-off production offers a number of benefits to the designer, but also some disadvantages.

Advantages include:

- The product can be customised to meet the exact requirements of the customer and designer.
- Work is carried out by skilled staff, so it is generally of a high quality.
- Flexibility is possible in the design, which can even be changed during production.
- Replacement parts can be made quickly.
- It can be used to make pre-production prototypes.

Disadvantages include:

- Production is more expensive.
- Extra time might be required to design and re-design during production to satisfy the exact customer or designer requirements.

Case study

Lifeboats are very specialist items and are manufactured using one-off production techniques at the RNLI's All-weather Lifeboat Centre (ALC) in Poole, England.

The Shannon class of lifeboat is the first modern all-weather lifeboat to be propelled by waterjets instead of traditional propellers, which improves manoeuvrability. Six of these lifeboats are produced at the ALC each year.

- It requires staff with specialist skills.
- It is often much slower than other methods of production.

Activity

In small groups, discuss the advantages and disadvantages of one-off production techniques for manufacturing the Shannon-class lifeboat.

Batch production

In **batch production**, products (or components of products) are manufactured in batches of a set quantity. Each separate manufacturing operation is carried out on the whole batch in one go. The next manufacturing operation is then carried out on the whole batch, and this sequence is repeated until manufacturing is completed. Figure 1.57 shows specialist electronic circuit boards, which are sometimes manufactured using batch production techniques.

Key term

Batch production Method used in manufacturing where products are made in a specific amount (a batch) within a specific time frame.

Figure 1.55 Shannon-class lifeboat

| Stage 1 Manufacture blank PCBs | Stage 2 Drill all PCBs | Stage 3 All components to all PCBs | Stage 4 Solder components into place on all PCBs | Stage 5 Final checking of PCBs and insert into product |

Figure 1.56 Batch production of circuit boards

Figure 1.57 Electronic circuit boards

Figure 1.56 shows the five steps the batch of circuit boards must go through before completion. Remember that they cannot move to the next step until all the circuit boards in the batch have completed the manufacturing operation at the current step.

Like one-off production, batch production has advantages and disadvantages.

Advantages include:

- There are lower production costs – machines and skilled staff can be used more efficiently.
- It offers the possibility of designing products with some variety.
- It is good for small production quantities and seasonal products.
- Consistent quality of products can be achieved.

Disadvantages include:

- Reconfiguring the production system for different batches takes time.
- If the design has an error, all products in the batch will have the same error.
- Costs are high because a stock of materials needs to be kept, ready to make products, and the finished products need to be stored before they are sold.
- Skilled staff could be demotivated over time.

Research

Using the internet, find other products that are manufactured using batch production techniques.

Mass production

Mass production is used to make large amounts of standardised products. It is sometimes called flow or continuous production. Mass production is used to manufacture single components such as fastenings (for example, screws and bolts) and products such as domestic appliances, cars and packaged food.

Key term

Mass production The production of a large quantity of a standardised product or component, often using automated production processes; also known as flow/continuous production.

Figure 1.58 shows the mass production of canned food. Mass-produced products made from assembled components are usually organised into production lines as shown. These are sometimes also called assembly lines. For complex products like a car, there may be many different assembly lines feeding sub-assemblies into the main production line.

Figure 1.58 Production line mass-producing canned food

Advantages of mass production include:

- It is economically very efficient, with lower production costs than one-off or batch production.
- There is a very fast rate of production – faster than any other production process.
- Product accuracy is high with **automation**, so fewer skilled staff are required.

Disadvantages include:

- Production lines are very expensive to set up.
- It is difficult to change the design once the production line has been set up.
- Staff might not be very motivated since their work is repetitive.
- If one part of the production line breaks, the whole process stops.

Key term

Automation Using computer technology to operate equipment, rather than humans.

Automation

Automation uses computer-controlled processes and machines to replace tasks usually undertaken by human beings. Examples of industrial automation include:

- computer numeric control (CNC) machines
- conveyors and moving systems
- robots
- automatic test equipment
- stock-control and order-processing systems and databases.

Figure 1.59 Industrial robots performing manufacturing operations automatically

Automation provides many benefits. It speeds up production time, with tasks and operations performed with fewer errors (that is, less human error). When an automatic machine is programmed to perform a task over and over, it has much better accuracy and repeatability compared to a human operating machinery. It is also generally safer as automation can perform operations that could not be safely carried out by humans (for example, in a hazardous environment). Fewer skilled staff are required to operate automated machines and processes. For the designer, it means that complex components and products can be manufactured – ones which would be difficult or impossible for a human to produce by hand.

There are some disadvantages to automation. It provides less flexibility as the machines and processes are generally programmed to undertake just one task. It is much more expensive to implement, with a large investment required at the start. Highly skilled staff are required to program and maintain automated systems. As automated machines are operated using motors, gases and fluids, they can increase the pollution caused by the manufacturing process.

Activity

1 Working in pairs, use the internet to find five different examples of how automation is used in mass production.
2 Copy and complete the table below, summarising the advantages and disadvantages of one-off, batch and mass production methods.

	Advantages	Disadvantages
One-off production		
Batch production		
Mass production		

New manufacturing processes

Improvements in materials and new manufacturing processes can be used together to optimise a design. New manufacturing processes can bring many advantages, such as faster turnaround times, reduced waste levels and improved product quality and performance.

However, designers have to be fully aware of the limitations associated with any manufacturing process, to ensure they develop a design suitable for that method. A new manufacturing process may allow development of a product with increased performance, but the cost of the process and the scale at which it can be used can make it unworkable.

Additive manufacturing, composite production and industrial automation are all examples of processes that can radically change how a design can perform. For example, additive-manufacturing processes can remove many of the limitations in geometry associated with traditional manufacturing methods. This gives the designer more freedom with the shape of

the design. However, if the product is to be mass produced, additive-manufacturing techniques may not be suitable (see page 58).

The designer may also consider optimising the assembly of the product, particularly if it is being mass produced. Automated production requires large amounts of money to purchase the necessary equipment, so designers need to consider the impact it may have on the final selling price based on the numbers that are produced. When designing components, the designer also has to consider how they will be held, processed and assembled on the production line. This may include considering the use of jigs and fixtures or how robots may move and hold components.

Stretch activity

Select a new manufacturing process from:

- 3D printing/additive manufacturing
- composite manufacturing
- automation.

List three advantages of the process you have selected. Are there any disadvantages?

Material availability and form

Selecting a suitable material for a design is more complex than checking that the properties of the material meet the product's performance requirements. The availability of the material also needs to be considered because **supply and demand** can change over time. If a product or component was designed and the material to produce it was not available, this could result in a delay in production or a modification to the design.

Key term

Supply and demand Relationship between the quantity of products a business has available to sell and the amount consumers want to buy.

A delay to production could mean that money made from sales of the product cannot cover the costs invested in design. Modifying the design to include a different material could mean a dramatic change of geometry, material properties or different manufacturing processes. To avoid this, material suppliers should be consulted early in the development of the design to ensure the materials will be available in the required quantities.

Many companies do not manufacture every element of a product on site. There may be many other businesses that provide components or resources required to manufacture the product. This could be because:

- the machines required are too expensive
- another business has greater expertise in a certain process
- it reduces the processes on site, allowing the business to produce finished products faster.

This network of businesses is called the supply chain. The supply chain also needs to be involved early in the design development to ensure it can provide what is required within acceptable timescales.

Some resources that can be provided by the supply chain include:

- raw materials
- components
- sub-assemblies
- tooling
- packaging
- **consumables** (for example, oil, lubricant and paint)
- distribution.

Key terms

Consumables Resources that assist manufacture and are used up during the process – for example, oil and lubricant used in machines.

Standard forms Made available in large quantities (often by mass production) to the same specification.

Most engineering materials are available in **standard forms** – that is, shapes, sizes and quantities to enable them to be easily used across a wide range of industries. The common forms of supply include:

- granules and pellets (such as aluminium, thermopolymers – for use in moulding and casting processes)
- ingots (usually metals that are cast into a shape that allows them to be more easily handled, stored, transported and processed)
- bar (such as aluminium, steels, thermopolymers – flat, round, hexagonal, in a wide range of sizes)
- sheet and plate (such as sheet metals, thermopolymers, manufactured boards, steel plate)
- pipe and tube (such as copper, steels, thermopolymers).

Types of manufacturing processes

A wide range of manufacturing processes can be used to produce components. A designer will decide which process is most suitable by considering:

- the materials the component is made of
- the geometry of the component
- the number of components that need to be produced.

The manufacturing process will have a big impact on the cost of production, which will affect the selling cost of the product. Designers need to make these decisions in collaboration with the manufacturer, to ensure the appropriate processes and associated considerations are set out in the design specification.

The major manufacturing methods are detailed below.

Wasting

Wasting is a process that uses tools to remove material from a workpiece until the required component has been produced; wasting can be achieved by hand tools and machine tools.

The hand tools used to remove material will depend upon the type of material and the shape of the required outcome: straight and curved cuts can be achieved using hand saws and surface smoothing and shaping can be achieved using hand files and chisels.

The main machining processes used to waste material are **turning**, **milling** and drilling. Each process can be carried out using manual or computer-controlled CNC machines, but the principles of removing material are the same in either case.

Turning uses a lathe to make round components. The material is held in a **chuck** on the lathe and rotated at speed. The cutting tool is then pushed into the workpiece to remove the desired amount of material and create the component. Figure 1.61 illustrates the various operations that can be undertaken using turning.

Key terms

Turning A machining operation that generates cylindrical and rounded forms with a stationary tool.

Milling A machining operation designed to cut or shape material using a rotating cutting tool.

Chuck Specialised clamp that holds material so it can be turned on a lathe.

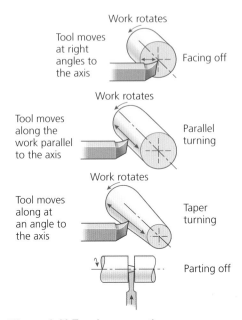

Figure 1.61 Turning operations

Milling uses a rotating tool to remove material. Slots, pockets, holes and flat faces can be produced on a milling machine by pushing the rotating tool into the material. Milling can produce detailed, accurate components with a high-quality surface finish. A typical vertical milling machine is shown in Figure 1.62.

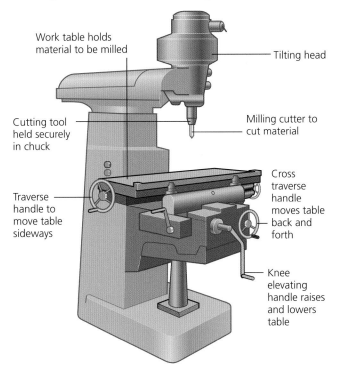

Figure 1.62 A vertical milling machine

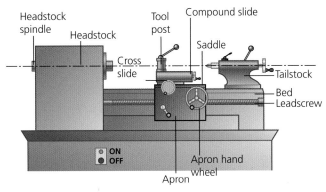

Figure 1.60 A centre lathe

Drilling is a wasting process that is used to produce round holes using a **drill bit** tool. The drill bit is held securely in the drill machine (such as the hand drill or pillar drill) by a chuck or **morse taper**, and as it rotates, it is driven into the workpiece. Large holes are usually produced in stages, first drilling a **pilot hole** and then increasing the size of the drill incrementally until the required hole diameter is achieved. There are many different types of drill bit, from multi-purpose drill bits through to specific drill bits for particular applications (such as twist drills for metal, brad point bits for wood, masonry bits for brick and concrete).

Figure 1.65 Different types of drill bits

Key terms

Drill bit A cutting tool used to create holes by removing material.

Morse taper A machine taper in the spindle of a machine tool or power tool.

Pilot hole A small hole drilled in material.

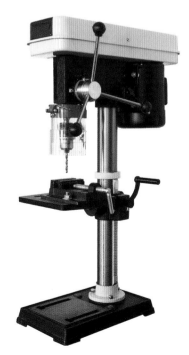

Figure 1.63 A hand drill

Figure 1.64 A pillar drill

Shaping

Shaping refers to processes that take raw materials and form them into final parts. Shaping processes include casting and moulding, machining processes such as turning and milling, and additive manufacturing. It also includes processes to shape material manually using hand tools, such as sawing and filing.

Figure 1.66 shows metal being shaped by casting. In metal casting, solid raw metal is heated until it is liquid and then poured into a mould of the correct shape and dimension of the component required. Once cooled and solid, the component is removed from the mould. This kind of mould can be made from different materials, such as sand or machined metal. Moulding can produce very accurate components which can be reproduced easily. This process can be used to produce one-off components and also a run of many identical components. Crucial to the moulding process is the design of the mould and the way the molten material behaves when filling the mould and once it has cooled. Moulded components may require further operations to prepare them after cooling, called **finishing operations** (see 'Finishing', on page 61).

Figure 1.66 Molten metal being poured into a mould

Activity

Find out what products are made by casting metals, and what different types of metal are used.

Stretch activity

Working in groups, research and summarise in a short presentation how each of these moulding processes works:

● sand casting
● investment casting
● die casting.

You should use simple diagrams to enhance your presentation.

Other material-shaping operations include those that can be carried out by hand. Manual operations like these are labour intensive and require skill to perform them safely and accurately. They are commonly used in the production of bespoke components and products, including prototypes. They are also used where the cost of machinery and automation cannot be justified for the quantity of components being manufactured.

It is exceptionally difficult, if not impossible, to produce complex components using manual shaping operations alone. For these types of

components, it is usual to perform materials shaping by machine, such as with computer numerical control (CNC), or by **computer-aided manufacture (CAM)**.

Figure 1.67 Filing metal by hand

Shaping by casting

Sand casting is used to make metal parts. The sand acts as the mould, packed around a pattern that is a representation of the component to be produced (Figure 1.68).

The sand sticks together because it is mixed with chemicals that allow it to bond. Once the sand has been packed around the pattern, the pattern is removed. Molten metal is poured through the **sprue**. Once the metal is solidified, the sand is shaken away, leaving the cast component. The sprue and **riser** are cut off and the part is finished.

Key terms

Finishing operations Operations carried out on a component to make physical corrections (for example, removing sharp edges) or to add a surface finish (for example, painting).

Computer-aided manufacture (CAM) Using computer software to control machine tools.

Sprue Passage created to pour molten material into a mould (the excess material that needs to be removed as a result of this process is also called a sprue).

Riser Vertical reservoir in the casting to allow molten metal to flow back into the mould cavity as the casting cools and shrinks.

Stage 1

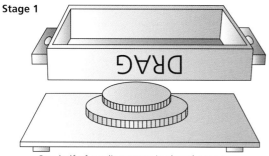

One half of a split pattern is placed onto a moulding board and the bottom half of the moulding box (the drag) is placed upside down over the pattern.

Stage 2 The pattern is cut through the centre and fitted with location dowels.

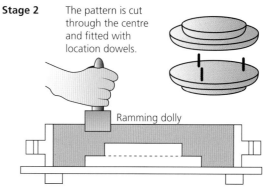

Ramming dolly

The pattern is sprinkled with a releasing agent (parting powder). Sand (Petrobond) is then added to the moulding box and rammed around the pattern.

Stage 3

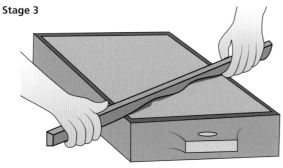

The top is levelled off (strickled) and the whole assembly turned over.

Stage 4 Sprue pins

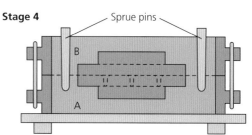

The top half of the moulding box (the cope) is then placed on top of the drag. The top half of the pattern is then fitted to the bottom half. Sprue pins are fitted, sprinkled with parting powder, and once again Petrobond sand is rammed around the pattern.

Stage 5 Pour metal via sprue

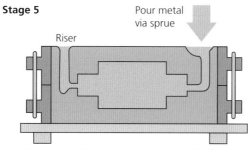

The pins are then removed, the top half of the mould taken off and the patterns are removed. Channels are then cut to connect the runner and riser to the mould cavity, and then the mould is reassembled and is ready for casting.

Figure 1.68 The sand-casting process

Die casting uses a special type of mould called a die. Molten metal is forced into the die under pressure, filling the cavity. Once the metal solidifies, the die is separated and the component can be removed.

Die-cast components tend to have a better surface finish and include more detailed features than those produced using sand casting. The die is made from high carbon steel, which is hard-wearing and has good shape memory, so it is mainly used to make parts from non-ferrous metal. The dies are usually expensive, so the process tends to be used when large numbers of components are going to be made.

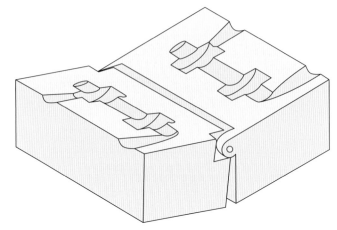

Figure 1.69 A steel mould used in die casting

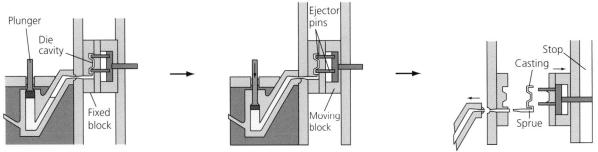

Hot-chamber high pressure die casting. A gooseneck hot chamber is submerged in a pot of molten metal. Metal is injected directly from the pot via the gooseneck.

Characteristic/properties	
Process	Permanent mould: Molten metal is forced into a water-cooled metal mould (die) through a system of sprues and runners. The metal solidifies rapidly and the casting is removed with its sprues and runners.
Shape	3D solid: Used for complex shapes and thin sections. Cores must be simple and retractable.
Materials	Light alloys: High fluidity requirement means materials with low melting temperature are usually used. Hot chamber method restricted to very low melting temperature alloys (e.g. Mg).
Cycle time	Solidification time is typically <1 second so cycle is controlled by time taken to fill mould and remove casting.
Quality	Good surface texture but turbulent mould filling produces high degree of internal porosity.
Flexibility	Tooling dedicated so limited by machine setting up time.
Materials utilisation	Near net shape process, but some scrap in sprues, runners and flash which can be directly recycled.
Operating cost	High, since machine and moulds are expensive.

Figure 1.70 Pressure die casting

Shaping by moulding

Moulding is the process of shaping a liquid or pliable (flexible) raw material using a mould. Examples include injection moulding, compression moulding and casting.

Figure 1.71 shows how injection moulding works. Raw material (typically plastic pellets but sometimes metal powder) is heated until it becomes liquid and then it is forced into a mould. The mould has been formed into the shape of the required component. Once cooled, the component is ejected from (pushed out of) the mould.

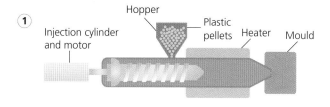

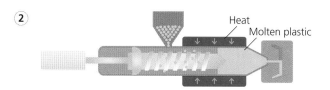

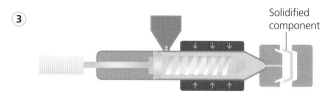

Figure 1.71 Injection-moulding process

Parts made by injection moulding must be carefully designed to allow the moulding process to be performed correctly. The properties of the materials used and the specific requirements of the moulding machine itself must be carefully considered. Despite these requirements, injection moulding can still produce accurate, low-cost components very quickly.

Activity

Write a description and produce a simple diagram explaining how injection moulding works and find out what products are typically made by injection moulding plastic.

Stretch activity

Working in pairs, investigate other types of moulding processes for plastics, how they work and what products they are used to manufacture.

Composite lay-up is a process of laying fibre and liquid polymer over a mould and allowing it to **cure**. Once cured, the composite can be separated and removed from the mould and it will be an inverse (opposite) replica of the mould. The mould side of the composite will be as smooth as the mould surface while the other side will usually be quite rough. The main composite lay-up processes are wet lay-up, spray lay-up and pre-impregnated lay-up.

- Wet lay-up, also known as hand lay-up, is a laminating process. A reinforcement material such as glass fibre is laid on an open mould coated with a primer and release agent. A polymer resin is applied and worked into the fibres of the reinforcement material using a brush or roller until it is soaked. Multiple layers of reinforcement material and resin can be added for greater thickness and strength. Once the required thickness is achieved, the lay-up is left to cure and harden before separation from the mould.

- Spray lay-up is a process of spraying a wet resin and chopped fibres mix on to a mould using a spray gun. Rollers are then used to even out and compress the fibre spray lay-up to remove bubbles and cavities.

- Pre-impregnated lay-up is a process used in the manufacture of dry flat-finish carbon-fibre components. Dry carbon-fibre sheet has resin pre-impregnated (pre-soaked) into the fibres. The carbon fibre is placed over a mould before pressure and heat are applied in an **autoclave** until the carbon fibre has cured and hardened.

Key terms

Cure To allow a lay-up to permanently take the shape of the mould over time (usually within a few hours).

Autoclave A sealed chamber in which products can be heated and pressurised under controlled conditions.

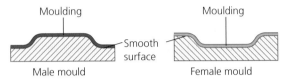

Figure 1.72 Male and female moulds for composite lay-up

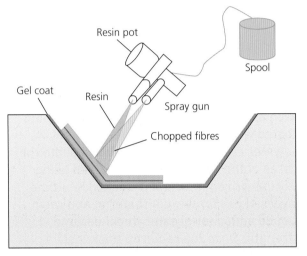

Figure 1.73 Spray composite lay-up

Shaping by machining

Machining describes a variety of material removal and shaping processes done by machine. It usually involves controlled removal of material using a cutting tool. This is sometimes called a **subtractive process** because material is being removed from a workpiece to produce the component or product, rather than building it up. Examples include turning, milling and drilling.

Key term

Subtractive process The removal of material from a solid block by a machining process, such as milling, turning or drilling.

Figure 1.74 shows material being shaped by milling. The milling cutter turns at high speed and removes material rapidly by moving into the workpiece.

The benefits and limitations of machining operations need to be considered during the design process.

Figure 1.74 A milling machine removing material, showing the waste material

Machining operations can be carried out at high speed, accurately, repeatably and safely. While machines can be controlled manually, they are often controlled by computer and automated. They make it possible to do things that would be impossible using hand processes, and a wider range of materials can be used.

Machines that can perform complex tasks are likely to be expensive to purchase, set up and maintain. That said, there is usually flexibility in the different types of component a machine can manufacture – with a computer-controlled machine, often reprogramming is all that is needed, although this does take time and skill.

Shaping by additive manufacturing

Unlike subtractive manufacturing, where material is shaped, moulded or removed, additive manufacturing is the process of joining material together layer by layer to create a component or product. Often referred to as 3D printing, it can use a range of different materials, including plastic, metal and concrete.

Additive manufacturing builds up 3D objects by adding layers of material under the control of a computer. CAD software is first used to design and create the component, which is then used to program and control the additive-manufacturing machine.

While this process has more commonly been used to quickly produce prototype components (rapid prototyping), it is now being used in the manufacture of actual components for use on finished products.

Case study

Components manufactured using additive-manufacturing techniques, like the ones in Figure 1.75, are cost-effective, as complex shapes can be produced more easily and they weigh less. It is also possible to customise designs. Airbus uses additive manufacturing for high-volume production of components for its aircraft and helicopters.

Figure 1.75 3D-printed parts

Research

Working in pairs, research how other manufacturers are using additive-manufacturing techniques when designing new products, and in the wider manufacture of products. Discuss your findings with the rest of the class.

Forming

Forming, pressing or press forming is where a pressing force is applied to a material to cause it to deform by bending or stretching and match the shape of a die (cavity). Once the pressure is removed, the material retains its shape.

Figure 1.76 shows metal being formed using a press tool. The machine performing the pressing operation is called a press brake. The metal is forced into the required shape using a die of the correct dimensions. If a different shape is required, a different die is used in the machine. The die being used is called the machine 'tooling' because it is the tool required to perform the operation.

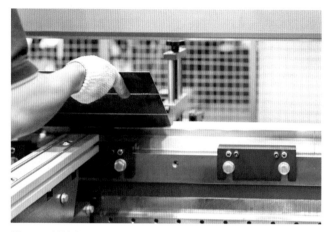

Figure 1.76 Press brake and tooling

Press forming is often used to shape sheet metal and plastics. Rolling, stretching, drawing and stamping are other examples of forming. The designer must consider carefully which forming processes might be useful in manufacturing their design.

Forming and pressing processes can be carried out at high speed and produce accurate and complex shaped components. They generally have lower production costs and require less-skilled staff to operate machines.

However, machines that perform these operations are expensive to buy and set up. Pressing and forming processes will almost certainly alter the properties of a material. Once set to perform a specific pressing or forming task, making any changes is costly and time-consuming, so it is best to get the design right first time.

Press forming is used to make components from metal sheet, such as car body panels. Complex press-forming processes may have multiple operations to create complex components.

Forming is explained further in Unit R040 (page 216).

Figure 1.78 Temporary fixings – nuts, bolts and washers

Figure 1.77 An industrial press and press tools

Joining

Joining processes are used to physically join two or more components. They can include joining methods such as soldering, brazing, welding, the use of adhesives and mechanical assembly, such as bolts, nuts and rivets.

Temporary joining methods allow components to be taken apart easily. These include mechanical fixings such as nuts and bolts, press-fit fastenings and self-tapping screws.

Permanent fixing methods, like those shown in Figure 1.79, are used when disassembly is not required and include fixings such as welding, brazing, soldering, riveting and the use of adhesives.

Figure 1.79 Using a rivet gun

Table 1.3 highlights some of the advantages and disadvantages of both types of fixing, which need to be considered at the design stage when selecting joining methods for assembly.

Activity

In small groups, discuss the benefits and disadvantages of using temporary and permanent fixings when designing new products.

Table 1.3 Advantages and disadvantages of temporary and permanent fixings

	Advantages	Disadvantages
Temporary fixings	• Components and products can be easily assembled and disassembled, requiring little skill and only simple tools • Can provide moderate force between the components being joined • Allow for slight movement between components	• Can loosen or fail with high levels of vibration – meaning components come apart • Can be affected by heavy shocks – breaking the fixing

	Advantages	Disadvantages
Permanent fixings	• Much more rigid compared to temporary fixings • Can withstand high levels of vibration and shock, so product does not come apart • Provide security from components and products being taken apart by the user or being vandalised	• Difficult to take apart for maintenance or repair • Need specialist skills to apply correctly, such as welding and soldering • Can be affected by environmental conditions • Faults in the joining process may lead to the fixing failing immediately or later on

Stretch activity

Working in pairs, copy and complete the table below to describe how various fixing methods work. You could illustrate your answers using simple sketches. Identify whether they are permanent or temporary fixings.

Fixing method	How it works	Permanent or temporary?
Nuts and bolts		
Welding		
Self-tapping screws		
Soldering		
Rivets		
Brazing		
Press fasteners		

Mechanical fixings are available in a range of standard sizes and specifications. Most are produced in large quantities by mass production to exact specifications and have been designed by experts in the field. They are called standard components and using them saves a significant amount of time and money in both the design and the manufacturing process. Lubricating oils, such as those used in a car engine, can also be considered a standard component, as can some batteries.

Resistors are one of the most commonly used electrical components. They are manufactured in large quantities in a range of preferred (or standard) values (Figure 1.80). Producing these to a custom value or specification is expensive and time-consuming. It is more efficient for the designer to select those with standard values in their design.

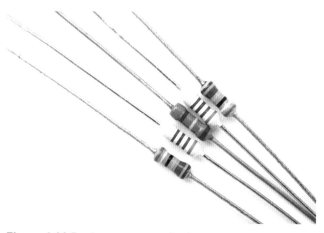

Figure 1.80 Resistors are standard components

Stretch activity

Investigate what range of standard sizes metric hexagonal nuts and bolts come in.

Key term

Lacquer Liquid that is applied and dries to form a hard, protective coating.

Finishing

Finishing refers to the process of applying a decorative appearance, protective coating or other treatment to a surface. Examples include painting, applying a **lacquer** and plating (for example, applying a chrome finish). **Heat treating** a surface to alter the properties of the material can also be considered a finishing process.

An alternative description of finishing is performing a finishing operation. Finishing operations are performed on a component or product to make it ready for the next stage. Examples include:

- removing spare (waste) material from a cast or moulded component once it has been removed from the mould
- cleaning a product before it is packaged up and sent to the customer.

In Figure 1.81, the sharp edges are being removed from a casting. This is called **deburring**. In this example, the finishing process is being done by hand, although it could be done using automation if the quantity being produced justified it.

Key terms

Heat treating Metalworking process where heat is applied to change the physical and sometimes chemical properties of a material.

Deburring Process to remove sharp or raised edges on a material caused by other processes, such as casting or machining.

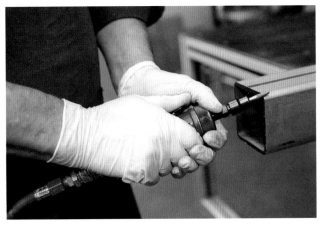

Figure 1.81 Deburring a casting by hand

Figure 1.82 shows a car being spray painted by robots. This is an example of finishing by automation. Painting not only improves the aesthetic appearance of the car, but it also provides protection against wear and tear and, more importantly, corrosion.

Figure 1.82 Robots spray painting a car

Assembly

Products with two or more components will have been designed for assembly and they will function because their components will accurately fit together when assembled. The components of bespoke products that have been manufactured as a one-off, or in very low numbers, will fit when assembled due to the craftsperson's high level of skills and ability. For batch and mass-produced products, mechanical devices such as jigs and fixtures will be designed to ensure that each component is manufactured or machined as accurately as required.

Assembly devices are used to securely hold a workpiece or component so that it can be accurately machined or assembled in a specific location or orientation. Multiple-use assembly devices are highly accurate and designed specifically for the task they perform.

For further information on assembly methods, see Unit R040, 'Physical modelling' (page 185).

Activity

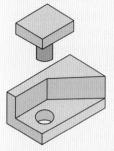

Figure 1.83 Designing a drilling and assembly jig

Design a drilling and assembly jig for the component shown in Figure 1.83:

- The jig must be able to securely locate and hold the lower component in a precise position for drilling.
- The jig must be able to guide the upper component to a precise position for assembly with the lower component.
- It must be possible to use the jig multiple times.

Production costs

Many factors affect the cost of production. By considering these costs early in the design process and detailing them in the specification, the designer can make decisions that will help them to manage costs when the product goes into production.

The following areas should be considered when calculating or managing production costs:

- material costs
- **tooling costs**
- machinery costs
- **labour costs**
- **overheads** – for example, power, transport, legislation.

All these costs need to be added together to give the total cost of production. If this total cost is then divided by the number of products to be manufactured, the product cost per unit can be determined.

The cost of the above categories can be influenced by the designer's decisions during development. For example, staff may be required to assemble a product. The more staff required, the higher the labour cost. The designer can consider how the product is assembled during its design to minimise the number of processes, make assembly as fast and simple as possible or consider using automation. This may require changes in material, geometry or components, so it is critical that design decisions are made at the appropriate time during product development.

The specification may detail critical product considerations that can influence the overall cost of production while also providing a maximum cost for production based on the expected selling price or budget.

Key terms

Tooling costs Cost of moulds, cutting tools, jigs or fixtures required to make a product.

Labour costs Cost associated with employees in a business, including wages, taxes and additional benefits.

Overheads Expenses that need to be paid by the business, not including labour or materials, such as rent and utilities.

Stretch activity

A company is introducing a new product to its current range. The table below shows the monthly costs for this product:

Type of cost	Amount (£)
Materials	28,000
Tooling	4,000
Labour	10,000
Overheads	5,000

1 The company predicts it will make 1,250 of these products in a month. Calculate the total monthly production costs and how much it will cost per unit.
2 The company plans to sell each product for £50. How much profit will they make from each product?

Labour

Labour costs are the sum of employee wages, taxes and any additional benefits that are paid by the employer. A typical engineering business will have direct and indirect labour costs. Direct labour costs are connected to the employees who are involved with the engineering side of the business, whereas indirect labour costs are for employees who help to maintain the function of the business, such as office staff, cleaners and drivers.

Capital costs

Capital costs are one-off expenses incurred when a business buys assets. Assets can be large and very expensive items, such as land, buildings and manufacturing equipment, or smaller items that will be replaced at regular intervals, such as chairs, desks, computers and vehicles. Capital costs also include non-physical (known as intangible) assets, such as licences and patents.

Test your knowledge

1 What are the main advantages and disadvantages of one-off production?
2 What is meant by the term batch production?
3 What are the standard forms of material supply?
4 What are the key differences between shaping by casting and shaping by machining?
5 What factors can impact on production costs?

2.3 Influences on engineering product design

Market forces are factors that affect the demand for new products – they can be complex and varied, and change over time. Companies, designers and manufacturers need to pay close attention to the forces influencing the market, to ensure products are developed in line with current or future demand.

New products are generally introduced onto the market because either:

- the market demands them (**market pull**), or
- new technology becomes available that drives their development (**technology push**).

Market pull

Market pull is when a need for a product develops from consumer demand – that is, people express a desire for a new product to solve an issue or problem they are facing. This could be through market research or feedback from users or consumers. It may also be in response to the introduction of new products from competitors, where a company sees an opportunity to develop a new product that can respond to this competition.

In the early days of the Covid-19 pandemic, there was market demand for face masks that could be used multiple times and that were aesthetically pleasing. Face masks have since become a multi-billion-pound market that is expected to continue to grow until 2027.

Technology push

Technology push is when investment in **research and development (R&D)** results in new technology being created that can lead to the development of new products. The technology created through R&D is built into new products, and then this 'pushes' products into the market. Technology push can occur with little or no market research and relies on demand being created by consumers being excited by the new technology.

Key terms

Market pull When a need for a product arises from consumer demand, which 'pulls' the development of the product.

Technology push When, as a result of research and development (R&D), new technology is created that can lead to the development of new products; products are 'pushed' into the market, with or without demand.

Research and development (R&D) Often the first stage in a development process, where companies carry out research activities to innovate and support the introduction of new products.

There are many examples of technology push, but one of the most well-known examples of recent years has been the introduction of touchscreen devices. As a result of R&D into touchscreen technology, Apple maximised the impact of 'pushing' this technology onto the market through the introduction of products such as the iPad. Previously, many consumers did not feel they needed an iPad in the same way that they needed a laptop, but the technology and design of the product excited the market and the iPad became a commercial success.

Activity

The Apple iPad and its integration of touchscreen technology is an example of technology push. Can you think of any other products that have been created as a result of technology push?

Cultural and fashion trends

Cultural and fashion trends regularly determine the direction of market forces that create demand for new products. In other words, many new products are developed in response to trends.

Throughout history, the aesthetics of products (see page 26) have been directly influenced by cultural and fashion trends. Many historical periods are now the subject of study by designers, who may look to previous trends for inspiration – for example, the Art Deco movement of the 1920s and 1930s or the rise of minimalist design from the 1960s to the present day.

Figure 1.84 Art Deco-style radio

Figure 1.85 Modern intelligent assistant smart speaker

Cultural and fashion-focused trends evolve and change over time. Companies invest large amounts of time and money to understand current trends and predict their future direction. This informs their design direction, so that they can gain a **competitive advantage** through greater understanding of **emerging markets**.

An understanding of cultural trends is essential if products are being sold across the world. Symbols, colours and product names may have different meanings in different countries, so careful consideration must be given to these elements of a design if trying to sell the product in other countries.

Key terms

Competitive advantage When a business is in a favourable position compared to other businesses because it has products, technology or market share that others do not have.

Emerging market Either a part of the world or a group of consumers that has been identified as potential future customers based on developing trends and behaviours.

A focus of modern design is consideration of a new product's impact on society and the environment. **Consumerism** has led to the development of a 'throwaway society', with large numbers of products flooding the market and being disposed of in a short period of time. Examples include low-cost fashion and single-use plastics. Consumers are becoming interested in more sustainable products, and designers should be aware of this when working on future products.

Key term

Consumerism A culture of excessive purchasing of material goods and services.

Figure 1.86 Social media is a recent cultural and fashion trend, with increased integration becoming more common in consumer products

Stretch activity

List a range of current cultural and fashion trends with which you are familiar. How are they influencing products? How do you think they might influence products in the future?

British and International Standards

Standards provide guidance on how to meet legislation. They give recommendations on how to design and produce products, and a set of minimum requirements.

Standards are written and agreed by experts based on their extensive knowledge of an area. They are regularly updated to reflect the latest legislation, accepted practices and advances in technology. They provide guidance to design

and manufacturing processes, but they are not design manuals and still need skill to be interpreted correctly.

Choosing to follow a standard means that a product is more likely to meet any legal requirements. Knowing that a product has been designed and manufactured to meet standards reassures the user and consumer. Manufacturers often include a list of the standards with which the product complies in their sales and marketing materials and manuals. The product might even have markings to show this.

Compliance with standards provides fairness for the manufacturer, as all manufacturers need to comply (meet the standards), and it also allows goods and products to be offered for sale in different countries. If all manufacturers of a particular product need to meet a minimum set of requirements, it means they cannot cut corners – for example, one manufacturer cannot deliberately produce an inferior (non-compliant) product to gain a market advantage over its competitors.

Activity

In small groups, discuss the relationship between legislation and standards. How are standards used to ensure that designs and products comply with legislation?

An example of a standard for the safety requirements of bicycles is *BS EN ISO 4210-2:2015: Cycles. Safety requirements for bicycles. Requirements for city and trekking, young adult, mountain and racing bicycles.* Designers and manufacturers need to check they are using the latest version of a standard, as they are updated regularly. This standard, although managed in the UK through the British Standards Institution (BSI) (see below), is an international standard as designated by the International Organization for Standardization (ISO).

While a single component may only need to comply with the requirements of one standard, more complex products will often have to comply with many. Both the design of the product and the actual, final product need to meet the requirements of standards, by following quality assurance procedures during manufacturing and regular inspections and testing when in use.

As well as product-specific standards, like the one for bicycles, there are also standards that relate directly to the quality of processes being performed. Quality standards provide a framework and guidance on how materials, products and services should be manufactured or delivered to ensure they:

- meet quality expectations
- are fit for purpose
- meet the needs of the user.

Quality standards are not usually product-specific and do not provide detailed guidance on designing the product, just on manufacturing processes.

The most commonly used quality standards are those in the ISO 9000 family. The ISO 14000 series of standards are also relevant, which concern how organisations can improve their environmental management efforts.

British Standards

British Standards are produced by the BSI. This is the UK national standards body, although as well as developing standards for use in the UK, it also works across the international standards community.

The BSI Kitemark is a quality mark owned and managed by BSI. It was created in 1903 and shows that a product or service meets the appropriate standards and that the quality management systems operated by a supplier are thorough. Products that comply with the BSI Kitemark scheme clearly display the mark shown in Figure 1.87. There are over 2,500 Kitemark licences held by manufacturers and suppliers.

Figure 1.87 The BSI Kitemark

Visit the BSI website (www.bsigroup.com/en-GB) for details of the range of standards the BSI produces and manages.

United Kingdom Conformity Assessment (UKCA)

Following the United Kingdom's departure from the European Union (EU), a new UK regulatory body, the United Kingdom Conformity Assessment (UKCA), has come into force for manufactured goods being placed on the UK market. It covers most products that previously required the European Conformity CE mark (see Figure 1.88). Products sold in the UK must show that they have been assessed by using the UKCA symbol. Products to be sold in Northern Ireland still need a CE marking to show they have met EU rules.

Figure 1.88 UKCA marking

Activity
Use the internet to research which standards are applicable to bicycle helmets.

European Conformity (CE)

European Conformity is certification that indicates conformity with the health, safety and environmental protection standards for products sold within the European Economic Area (EEA) – that is, the EU and several connected countries. The CE marking is also recognised worldwide.

Figure 1.89 CE marking

By displaying the CE marking on a product, a manufacturer is declaring that it meets the requirements of any applicable EU directives or regulations. As well as having to meet these requirements, the manufacturer also needs to complete a declaration document. Sometimes this document is included with the printed material (instruction manual) provided with new products. Often it is just available on the manufacturer's website or on request. The declaration includes:

- the manufacturer's details
- essential product characteristics
- the European standards with which the product complies
- performance data
- a signature on behalf of the manufacturer.

The manufacturer can place the CE marking on its product personally, if it is confident about compliance. However, it is more likely that an outside organisation will need to check and test the product to confirm that it complies.

Activity

Identify at least five products that have the CE marking on them.

When we use a product, we want to be sure it will work safely and reliably, perform as intended and not cause harm to either us or the environment.

We can have confidence that products designed to conform with current government legislation and appropriate standards will work reliably, safely and with environmental responsibility.

Legislation

Legislation means laws proposed by the government and made official by Acts of Parliament. If an individual or company can be shown to have broken a law, they can be prosecuted, with severe consequences, such as a fine or a prison sentence.

There are many pieces of UK legislation that are relevant for businesses, such as the Health and Safety at Work Act 1974 (HASAWA). This lists a wide range of duties of both employers and employees to ensure 'health, safety and welfare' in the workplace.

The Health and Safety (Enforcing Authority) Regulations 1998 allocates inspection and enforcement of health and safety law to the Health and Safety Executive (HSE) and local authorities (LA) according to the main work activity being undertaken.

A **risk assessment** is an example of HSE regulations in the workplace. All potential hazards must be risk assessed so that risk of injury or damage is minimised to an acceptable level.

In a risk assessment, all hazards presented by an activity are identified, together with the risks they present and the **control measures** that must be followed before the activity commences. For further information on this and how to write a risk assessment, see Unit R040, page 168.

Key terms

Risk assessment Process of identifying, analysing and evaluating hazards and their associated risks, and seeing if the risks are acceptable or can be reduced.

Control measures Actions taken to reduce the risk (likelihood of a hazard causing harm), such as removing the hazard, taking extra care, guarding or wearing protective equipment.

In order to trade with markets outside of the UK, businesses must ensure that products made in the UK comply with relevant legislation and standards for those markets.

In the EU, the terms 'regulation' and 'directive' are used to describe a legal act with which its member states (each country) need to comply. For example, the Machinery Directive is designed to ensure a common safety level in machinery placed on the market or put in service in the EU.

Legislation protects the consumer or user. Everyone needs to comply with it – such as when employing staff in the workplace or manufacturing a machine.

However, legislation can be difficult to read and understand. For this reason, manufacturing businesses often have access to experts who can help them interpret legislation when required.

The Waste Electrical and Electronic Equipment (WEEE) Directive is an example of legislation. It is an EU directive covering the collection, recycling and recovery of waste electrical and electronic goods. While complying with it is a legal requirement, each EU member state can decide how best to comply. The UK was an EU member state when the directive was adopted (2013), so it became law in the UK (2014).

The symbol used to represent waste electrical and electronic equipment is shown in Figure 1.90.

Figure 1.90 WEEE marking

Before the WEEE Directive, waste electrical and electronic equipment in the UK was often disposed of and processed alongside other household waste. Since the WEEE Directive, waste electrical equipment can still be taken to designated waste-recycling centres, but it is then sent on to specialist recovery and treatment facilities where it can be recycled or disposed of safely.

Right to Repair

High-profile brands and manufacturers have increasingly made their products difficult to repair by non-specialised service centres, and they have restricted the availability of replacement components. In 2021, the EU introduced a directive that promotes the repairability of a range of domestic products, such as dishwashers, washing machines, refrigerators and televisions. For new products to be sold in the EU, they must be designed and manufactured so they are able to be repaired by non-specialised repairers and components for repair must be available for up to ten years. The UK Government is planning to introduce similar legislation.

Research

Research 'Right to Repair' and the effect that the legislation could have on the lifespan of domestic products such as dishwashers, washing machines, refrigerators and televisions.

Planned obsolescence

All manufactured products have a lifespan, a period of time when they should function as intended without failure. If a product has moving parts, the chances are one or more of these parts will wear over time due to repeated use; if it has electronic components, again, over time one or more may fail. UK law requires manufacturers and retailers to be responsible for ensuring that certain types of consumer products safely fulfil their function for a guarantee period, usually up to one year after purchase. Many retailers also offer service and repair facilities so that products can be maintained in good working order.

Over time, designers and manufacturers modify and update their products in line with changes to trends and fashion, new materials and new technologies, and older products become **obsolete**.

Planned obsolescence is the strategy of planning a deliberately shorter lifespan for a product. This can be seen in a number of ways:

- Throw-away products – this requires the consumer to consider a product not to be worth the cost of repairing it. The product may contain lower quality components that are expected to fail after a limited lifespan, and service, repair and replacement of the components may be greater than the cost of a new product.

- Software updates – products that require software for them to function may only be available on a certain model for a limited period of time. Updates to software may not offer the full range of serviceability, connectivity to **peripheral devices** or speed.

- Restricted serviceability – with this strategy, access to branded components is restricted to officially licensed service agents. As a result, maintenance and repair of products can be a slow and costly process; rather than wait for a repair, customers may be encouraged to replace a defective product with a new one.

- Service agreement obsolescence – products can also be linked to service agreements that have a pre-determined lifespan, at the end of which the customer is offered an upgrade to a newer model on preferred terms. Towards the end of service agreements, high-profile brands may use targeted **perceived obsolescence** marketing strategies to introduce model upgrades with subtle additional features; the new models become the 'must-have' product, while the old product appears dated even when it may be only one or two years old.

Sustainable design (6Rs)

All products eventually come to the end of their life. This could be when they have been used up or are broken.

The design of a product can influence what happens to it when it reaches the end of its life. This includes the materials and assembly methods selected to manufacture the product, as well as packaging materials.

The **6Rs** provide a useful set of questions to assess the sustainability of a product. A designer can continually refer to the 6Rs to check what impact a product may be having at each stage of its development.

Key terms

Obsolete When a product becomes outdated or unserviceable and parts are no longer available to repair them.

Planned obsolescence A strategy of planning a deliberately shorter lifespan for a product.

Peripheral devices Devices that connect to a host product, such as a watch, speaker or earphones (peripherals) connecting to a mobile phone (host).

Perceived obsolescence A marketing strategy that encourages consumers to upgrade to the latest model.

6Rs Areas to be considered when assessing the sustainability of a product: recycle, reuse, repair, refuse, reduce and rethink.

Table 1.4 The 6Rs

Recycle	• Can recycled materials be utilised to manufacture the product? • Can any waste material from production be recycled? • Can the product be easily disassembled so it can be recycled? • Can the materials be recycled at the end of the product's life?
Reuse	• Can the product or its parts be reused for another purpose at the end of its life? • Can the product be used multiple times rather than single use?
Repair	• Can the product be designed to allow it to be repaired instead of thrown away?
Refuse	• Can the designer refuse to use materials that are not recyclable? • Can the designer or manufacturer refuse to use non-renewable energy sources?
Reduce	• Can the energy used in manufacturing or operation be reduced? • Can the amount of waste during production be reduced? • Can the product be made from fewer materials or can the amount of unsustainable material be reduced?
Rethink	• Can the designer rethink the way the product is designed or manufactured? • Is the material being used the only option?

Activity

Assess a product of your choice using the 6Rs. Can you identify any ways that the product could be improved?

Use of materials at end of life

The sustainability of a design can be dramatically improved if the materials used in the product can be separated at the end of its life and recycled, reused or upcycled. There are many examples, particularly where materials are upcycled, of creative solutions to reuse materials in new products, avoiding the need to dispose of them in landfill.

Recycling materials

Recycling is the process of converting waste materials and objects into new materials and objects. It not only saves limited and precious natural resources, but it also reduces harm to the environment.

People are now very conscious of the need to recycle products and packaging. Each local authority in the UK has arrangements to recycle household waste. We are even encouraged to separate waste into categories – like plastics, glass and general waste. We are discouraged from placing certain items in our general waste, such as electrical equipment and hazardous products like batteries, which need to be taken to a local recycling point or centre for specialist recycling or disposal.

Engineered products are manufactured using a range of materials. Some can be easily recycled and others cannot. Materials that are easier to recycle include metals like aluminium, iron and steel and other materials such as glass. On the other hand, certain plastics are less easy and sometimes impossible to recycle. There are different procedures for recycling larger and complex products such as electrical equipment.

Assembly and construction methods affect the potential to recycle a product. If the product can be easily broken down into its basic components, and similar materials can be grouped together, then recycling is much more straightforward. A simple example of this is removing the metal lid from a glass jar – both metal and glass are recycled separately. Packaging materials should also be recyclable, which often happens long before the end of a product's life.

Materials that cannot be recycled often end up in landfill sites or are **incinerated**, potentially releasing harmful gases.

Key term

Incinerated When something is burned to dispose of it.

Figure 1.91 Waste being disposed of in a landfill site

To reduce waste going to landfill, designers need to consider carefully the materials and manufacturing processes, including assembly methods, used when creating products. They also need to decide how the product will be packaged for safe storage and transportation. The sustainability of materials is increasingly becoming more of a consideration for consumers when they buy products.

Products often include symbols to show if they can be recycled or not. Figure 1.92 shows a recycling symbol for a product containing aluminium.

Figure 1.92 Aluminium recycling symbol

Research

Find out what other recycling markings and symbols are used on products.

Case study

Recycling car tyres is challenging due to the large volume manufactured and the rubber materials and compounds they are manufactured from. Because tyres are bulky, durable and non-biodegradable, it is best to avoid putting them into landfill.

However, waste tyres can be reprocessed into products such as chipping that can be used as flooring in children's playgrounds (Figure 1.93).

Figure 1.93 Recycled tyres being used as playground flooring

Reusing products and components

Reuse is different from recycling. Reusing something does not require extensive reprocessing of previously used products, helping to save time, money, energy and resources.

A toner cartridge from a laser printer is an example of a product that can be reused because it can be refilled to allow it to be used again. The resources that have been put into manufacturing the toner cartridge from new have not been lost, so the refilled cartridge is much cheaper to purchase and better for the environment.

There are numerous benefits to reusing components and products but also some disadvantages. These are summarised in Table 1.5.

Key term

Reuse Practice of using something again for its original purpose or to fulfil another function.

Table 1.5 The advantages and disadvantages of reusing products and components

Advantages of reuse	• Raw material is saved. • The energy from original production is not wasted. • It reduces the need to recycle. • It reduces the need to dispose of waste material that is non-recyclable. • Refurbishment can create jobs, including those for developing economies. • It is cost-saving for the consumer.
Disadvantages of reuse	• It requires transport and so has environmental costs. • The product might contain hazardous materials or be less energy efficient than newer/current products. • Sorting and preparing items takes time. • Specialist skills might be required to prepare or repurpose an item. • The item might not comply with current practices and standards. • The item's appearance might have **deteriorated** – for example, due to corrosion.

Key terms

Deteriorate Get worse with age or fall apart.

Robust Strong, hard-wearing, less likely to break.

Remanufactured Rebuilding of a product to its original specification through repair or the use of new parts.

It is possible to design products to be reusable in the first place. The design requires more thought and often the original product needs to be more **robust**. This could mean that it is more expensive to manufacture and purchase. These are called products that can be **remanufactured**. Products that are commonly remanufactured include engines, office furniture, cameras and aircraft fuselage. Remanufacturing is often carried out by the original manufacturer or by specialist organisations.

Products can also be reused with little or no reprocessing or remanufacturing. There are lots of organisations, many of which are social enterprises and charities, that collect, refurbish and supply products for reuse. This can also create jobs. Some of these organisations support communities in countries with developing economies.

Upcycling

Upcycling is the process of finding creative uses for products at the end of their life, so that they do not need to be recycled or disposed of. Examples of upcycling include using glass or plastic products as decorative vases or plant pots, or giving old furniture a new finish to appeal to new markets. Figure 1.94 shows an example of upcycling, where wooden pallets have been used to create modern outdoor seating.

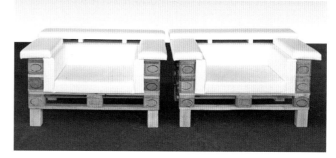

Figure 1.94 Upcycling wooden pallets to create contemporary garden furniture

It may appear that upcycled products are only created by individuals rather than by large-scale manufacturers. However, well-known international brands are increasingly finding innovative ways to upcycle material.

In 2016, Adidas began the production of shoes manufactured from plastic waste recovered from beaches and coastal regions. Their target was to produce 15 to 20 million pairs of shoes containing recycled plastic waste by 2020, a quantity that represents over 3,000 tons of plastic that would have been saved from the ocean. Adidas also use ocean plastic in the manufacture of sports clothing. Their 2024 target is to use only recycled polyester in every product and on every application where a solution exists.

Activity

Do you have any creative ideas for items that you could upcycle? Try to create a design using upcycled materials or products.

Disposal of non-recyclable materials

Despite large amounts of material being recyclable, a substantial amount of material still needs to be disposed of at the end of its life because it cannot be reused or processed into something else. The final destination of much of this material has traditionally been landfill, which involves taking waste and burying it in the ground. This can cause a range of environmental problems, such as contaminated soil or water pollution, and many landfill sites have become overloaded with waste.

The landfill problem has led to a series of companies trying to find solutions to the disposal of non-recyclable materials:

- Terracycle specialises in the processing of waste that traditionally cannot be recycled. They do this in many ways, including reusing, upcycling and exploring innovative solutions to recycle apparently unrecyclable products.
- In Sheffield, a waste collection company runs the 'District Energy Network'. Non-recyclable waste is collected and incinerated, and the heat is then used to feed heating and hot water across the city to more than 140 buildings. The buildings using the system need no additional fuel to heat them.

Sustainable engineering initiatives

Creating a sustainable future is a priority for many businesses and international organisations. Governments have worked alongside industry to create a range of initiatives, with the aim of improving sustainability across the world.

In 2016, world leaders agreed to accelerate the action and investment required to realise a sustainable, low-carbon future through the Paris Agreement.

In 2020, following consultation with stakeholders, industry and the wider public, the UK

Government announced the end of the sale of new petrol and diesel cars in the UK by 2030, with all new cars and vans to be fully zero emissions at the tailpipe from 2035.

At COP26 (2021), the UK Prime Minister launched a 'Clean Green Initiative' to help developing countries take advantage of green technology and grow their economies sustainably with a major funding package of over £3 billion to support the rollout of sustainable infrastructure and revolutionary green technology in developing countries.

Stretch activity

Research sustainable engineering initiatives. Can you find any other initiatives that have been undertaken by governments or businesses to improve sustainability? What did the initiative do and what has been the impact? Create a presentation to deliver to your class.

In recent years there has been increased concern about world-wide climate change, the dangers it presents and its causes. In response to the Kyoto Protocol (1997), the Paris Climate Accords (2015) and COP 26 (2021), national governments have made ambitious targets to reduce greenhouse gas emissions. To achieve international aims, industry and consumers around the world will need to change their perspective when it comes to everyday products. New industries have been created with dedicated aims for recycling and reusing materials, components and products.

Due to the availability of new products offering additional, faster, more powerful and longer-lasting features at relatively low-cost and a **consumerist** market, many consumers now discard materials and products long before they have become obsolete.

Key term

Consumerist Having a desire to own consumer goods and material possessions

Repair

An alternative to renewing a product when it fails to function as intended is repairing it or having it repaired. Large expensive appliances, such as washing machines, dryers, cookers and fridges have for many years been designed for manufacturing, assembly and disassembly so that they can be repaired. If they have moving parts, lights or heating elements, these are considered consumable items that will require replacement after a reasonable period of time. The unscrewing of fastenings and the removal of a casing or panel will expose the components so that they can be replaced.

Unfortunately, many smaller products are manufactured in a way that makes it difficult for consumers to repair them themselves or even have them repaired. This may be due to components not being accessible due to sealed units, such as casings that cannot be opened without damaging the product, or not making components available to third party repairers. Many large businesses have in-house repair centres and do not allow other repairers access to the components of their products. They can set a high price on the cost of a repair because they have a monopoly on who can repair the product. A number of governments have introduced 'Right to Repair' legislation that provides consumers with the freedom to choose where they have a product repaired: in such countries businesses must make original equipment manufacturer (OEM) components available at a fair price to repairers outside of their business. Some examples of these are OEM batteries, screens and printed circuit boards.

Reduce

Designers, manufacturers, retailers and consumers can all play a part in reducing greenhouse gas emissions. For example, designers and manufacturers could bring fewer products to the marketplace by reducing the range of models available and future-proofing products so that they can be upgraded. They could reduce the materials used in their products, reduce the amount of energy they use to make them and reduce the amount of energy their products consume. Retailers could reduce the

miles that products travel to and from their shops by not holding stock in shops. Their sales could be linked to direct delivery to the consumer from their warehouses or the manufacturer. Consumers could reduce the number of products they buy and only buy when absolutely necessary. They could demand products that consume low amounts of energy and are packaged only in essential packaging that must be 100 per cent recyclable, such as paper carrier bags instead of plastic carrier bags or card packaging instead of polystyrene.

Figure 1.95 A consumer picking the amount required rather than buying pre-packed

Refuse

Bringing a product to the marketplace can be very expensive, so designers and manufacturers seek to produce products that they believe will sell. To help achieve this they will seek consumer views through market research in order to know what the consumer desires from a product.

When a consumer buys a product, they may have already refused an alternative product. The criteria used for the purchase may have been inspired by the huge marketing budget of a major brand, or simply its technical performance. Consumers must be encouraged to consider their need against issues that are more rooted in environmental and sustainability concerns than fashion and trends: products that are inefficient or made from resources damaging to the environment should be refused. These include products that use too much energy and are too expensive to run or maintain.

In 2014, over 7.6 billion single-use carrier bags were given to customers by major supermarkets in England. In an effort to limit the number of single-use carrier bags, in 2015 the UK Government introduced a 5 pence levy on single-use carrier bags sold or given away by larger retailers. By 2020, major supermarkets supplied just 564 million single-use carrier bags. In 2021 the levy was raised to 10 pence per single-use carrier bag in England for all retailers. This is an example of a government using legislation to encourage consumers to refuse a product.

Rethink

'Do I really need it?' and 'Do I really need that one?' are questions that a consumer could ask themselves prior to the purchase of a new product. Rethink is about encouraging consumers to break away from consumerism and the need for more and more products that offer very little more than the products they already have. For example, many mobile phones are bought on service contracts with upgrades to a new model available after an agreed period of time. Consumers are encouraged to upgrade even when their older existing mobile phone functions perfectly well. Rethink is about questioning if the upgrade is really necessary. As well as upgrading to a new phone, many consumers often also upgrade their mobile phone accessories. In 2020, the UK smartphone accessories market was worth £1.9 billion servicing 65 million mobile internet users. By declining to regularly upgrade, consumers could send a message to manufacturers to rethink their business models to ones that are more environmentally friendly.

The motor industry has responded well to pressure groups and government legislation and has been forced to rethink the life cycle of a motor vehicle. For many years cars were manufactured from materials that were harmful and drained natural resources, but initiatives such as the UK End of Life Vehicles Regulations (2015) banned the use of materials such as cadmium, lead, mercury or hexavalent chromium, and set high annual targets for material recovery (95 per cent) and recycling (85 per cent by average weight of

each end-of-life-vehicle). Motor manufacturers have demonstrated how industry can, if pushed, rethink the design and production of products.

Packaging design is another example where designers and manufacturers were required to rethink to reduce the amount of waste and impact of packaging. UK regulations first launched in 1997 forced businesses using packaging to take responsibility for its environmental impact. All businesses that produced or used packaging were required to contribute towards the cost of recycling and recovery. This initiative aimed to reduce the amount of packaging that ends up in **landfill**.

Activity

Identify several consumer products that could perform the same function were they to be manufactured from a more environmentally friendly material.

For example, drinking straws can be made from paper rather than plastic.

Design for the circular economy

Traditional design values were based on products that ultimately were not designed to last. In contrast, **design for the circular economy** seeks to change design thinking from a 'take-make-waste' model to a model grounded on reuse, repair, remanufacture, refurbish and recycle, minimising the use of **primary raw materials**.

The circular economy is based on three principles:

- design out waste and pollution
- keep products and materials in use
- regenerate natural systems.

Key terms

Landfill A tip, rubbish dump or site for the disposal of waste materials

Design for the circular economy Economy model based on economic growth without consumption of finite resources.

Primary raw materials Naturally occurring substances extracted from the earth.

Economic growth has been dependent on a **linear economy** model of taking natural resources and turning them into products that are ultimately destined for waste and pollution. It has been driven by an unsustainable demand for new and improved products without consideration of their impact, including the reduction of natural resources and the effects of waste and pollution. In contrast, design for the circular economy is a model based on the natural world that takes what is needed to sustain life and then returns it pollution free for regeneration at the end of life. It is a model that promotes the use of resources for much longer before they are reused, repurposed or recycled without waste or pollution: economic growth without depletion of finite (limited) resources.

Key term

Linear economy Economy model based on the extraction of natural resources for products that will eventually end up as waste and potential pollutants.

Case study

The Ellen MacArthur Foundation develops and promotes the idea of a circular economy. It works with businesses, policymakers and institutions, highlighting the benefits of a circular economy and providing support and examples of how it can be put into practice (https://ellenmacarthurfoundation.org/).

Test your knowledge

1. What is meant by the term technology push?
2. Why might a business want its products to display the BSI Kitemark?
3. Select one piece of UK legislation that is relevant for businesses. Who might be affected by this legislation?
4. What do you understand by the following terms?
   - refuse
   - rethink
   - upcycling.
5. What is a principle on which the circular economy is based?

Practice questions

1. Give one example of a 'wasting' manufacturing process. [1]
2. Identify the criteria which best describe the way a product looks. [1]
3. Give two reasons why jigs and fixtures are used in assembly. [2]
4. Explain what is meant by the term planned obsolescence. [4]
5. Discuss what is meant by the term design for the circular economy. [6]

Getting started

Think about the different types of images and diagrams you encounter each day, and the information they communicate to you – for example, a fire exit sign in a room or a no entry road sign.

In pairs, discuss these different images and diagrams and what makes the information they communicate easy to understand and accurate.

3.1 Types of drawing used in engineering

Designers and engineers use a wide range of different methods to communicate and exchange ideas. The use of drawings and diagrams is the language of the designer – to communicate their ideas so that they can be reviewed and commented on, and to enable them to be manufactured into physical products. Each type of drawing has its own advantages and disadvantages. In this section, we will look at different types of drawings and diagrams used by the designer and engineer. In Units R039 and R040, you will have the opportunity to develop your knowledge further and practise engineering drawings and the use of computer-aided design (CAD) in much more detail.

Freehand sketching

Freehand 2D and 3D sketches are often used by the designer to quickly generate a range of different design ideas. They are quick and cost-effective to produce, can be easily shared with clients and users to gain feedback and do not require expensive or time-consuming modelling software. Producing good sketches requires some initial practice to develop the sketching skills required. See Unit R039, Topic area 1 (page 105) for more details on the different techniques for producing 2D and 3D sketches.

Isometric and oblique drawings

Isometric and oblique drawings (Figure 1.96) are 3D pictorial drawings used to show different views of a component or product. With **isometric drawing**, the focus is on the edge of the object, while with **oblique drawing**, the focus is on a face. Isometric drawings are drawn at an angle of 30° to the horizontal, and oblique drawings at an angle of 45° to the horizontal. Although isometric and oblique views are useful for providing a visual (pictorial) representation of different aspects of an object, they are often not detailed enough for manufacturing purposes. See Unit R039, Topic area 2 (page 121) for more details about how to draw isometric and oblique drawings.

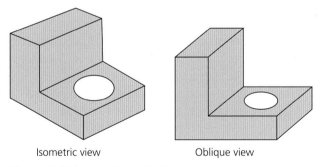

Isometric view · Oblique view

Figure 1.96 Isometric and oblique views

Key terms

Isometric drawing 3D pictorial drawing that focuses on the edge of an object and uses an angle of 30° to the horizontal.

Oblique drawing 3D pictorial drawing that focuses on the face of an object and uses an angle of 45° to the horizontal.

Orthographic drawings

An **orthographic drawing** represents a 3D object by using several 2D views of it. Typically, an orthographic drawing shows three different views, as shown in Figure 1.97, from the front, side and top of the object. It is a technical drawing that defines very accurately the requirements of the component, enabling it to be successfully manufactured. Being able to produce or interpret orthographic drawings correctly does require some skill and practice, as engineering drawings are drawn to national and international standards. Orthographic drawing is explained in more detail in section 3.2 of this unit (page 82), and in Unit R039, Topic area 2 (page 121), where you will have the opportunity to practise this type of drawing.

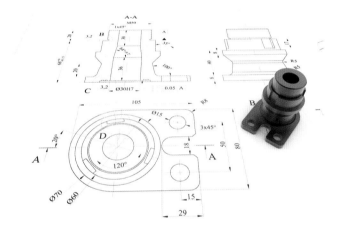

Figure 1.97 Orthographic drawing

Exploded view drawings

An **exploded view drawing** shows the component parts of a product separated and moved outwards. It shows the relationship of components to each other and is useful to see the order in which they are assembled to make the final product. Figure 1.98 shows an exploded view of a battery-operated toy car. Note that the components are numbered and a list of parts is included. While exploded view drawings show clearly how components fit together, they often do not provide sufficient detail to manufacture each component. This requires a separate orthographic drawing of each component. See Topic area 2 in Unit R039 (page 121) for more detail.

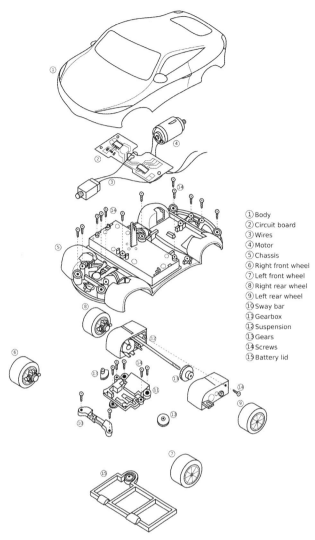

① Body
② Circuit board
③ Wires
④ Motor
⑤ Chassis
⑥ Right front wheel
⑦ Left front wheel
⑧ Right rear wheel
⑨ Left rear wheel
⑩ Sway bar
⑪ Gearbox
⑫ Suspension
⑬ Gears
⑭ Screws
⑮ Battery lid

Figure 1.98 Exploded view drawing

Key terms

Orthographic drawing Drawing that represents a 3D object by using several 2D views (or projections) of it.

Exploded view drawing Drawing that shows the component parts that make up the product separated and moved outwards.

Assembly drawings

Assembly drawings (Figure 1.99) show how multiple components or parts are assembled to make a final product. They show the parts in the correct position and how they are assembled (or joined) to each other. Assembly drawings can be either 2D (as shown) or 3D and show clearly how separate components fit together. Like exploded view drawings, they often do not show sufficient detail to be able to manufacture the separate component, so each requires its own orthographic drawing. See Topic area 2 in Unit R039 (page 121) for more detail.

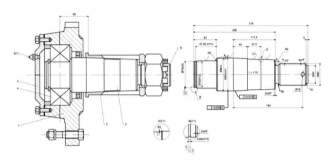

Figure 1.99 Assembly drawing

Block diagrams

A **block diagram** uses simple blocks, lines and arrows to represent as a diagram a more complex system. Figure 1.100 shows a simple block diagram of a washing machine controlled by a microcomputer. Note the input devices on the left and output devices on the right of the diagram, as well as the arrows showing the flow of information (or signals). Block diagrams are useful to understand simply how a system or process works. However, they do not show in detail the technical detail and workings of the content of each block in the diagram.

Circuit diagrams

A **circuit diagram** is a particular type of diagram used in electrical and electronic engineering. It uses commonly understood symbols of electronic components and shows how they are interconnected to form a working circuit design. Figure 1.101 shows a simple timer circuit that is light activated. Circuit

diagrams provide an easy-to-follow schematic layout of the circuit. However, they do not represent what the physical circuit looks like and it takes some practice to learn what each of the circuit symbols represents.

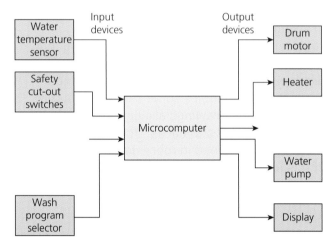

Figure 1.100 Block diagram

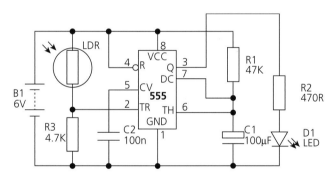

Figure 1.101 Circuit diagram

Key terms

Assembly drawing Drawing showing how separate components fit together.

Block diagram Diagram that shows in schematic form the general arrangement of the parts or components of a complex system or process, such as an industrial apparatus or an electronic circuit.

Circuit diagram Graphical representations of electric circuits, where separate electrical components are connected to one another.

Flowcharts

A **flowchart** is a type of block diagram where blocks are joined together using lines and arrows to show how various processes are linked together to achieve a specific outcome. They can be used for planning purposes and also for things like writing computer programs. Flowcharts are useful for showing visually how a process works using a series of different types of symbols to represent functions and decisions. It does, however, take time and practice to learn which symbols to use and to be able to convert the flowchart into a working system or computer program. You will find more details of the different symbols used in Topic area 2 of Unit R040 where a flowchart is one method of planning for making a prototype (page 189).

Wiring diagrams

Like circuit diagrams, **wiring diagrams** are used by electrical and electronic engineers to show the layout and wiring of a system. They show how plugs and sockets, and different coloured electrical wires and cables are used to interconnect electrical devices. Like circuit diagrams, they are a visual representation of how a system is wired, which makes them easy to follow. However, they do not show what the physical layout of the system and wiring looks like, just which wire connects to where. Figure 1.102 shows a wiring diagram for a motorbike – note how the headlight and taillight are wired into the system.

Key terms

Flowchart Block diagram that shows how various processes are linked together to achieve a specific outcome.

Wiring diagram A simple visual representation of the physical connections and physical layout of an electrical system or circuit.

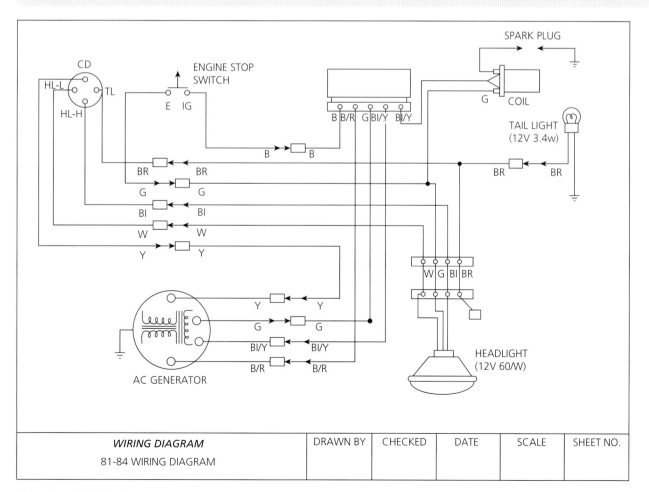

Figure 1.102 Wiring diagram

Activity

Draw a table like the one below and summarise the key characteristics, advantages and disadvantages of each of the different ways to communicate information. One of these has been done for you.

Type	Key characteristics	Advantages	Disadvantages
Freehand sketching			
Isometric drawing	• 3D pictorial drawing • Drawn at an angle of 30° to the horizontal • Focuses on the edge of the object	• Good for visual representation of the edge of an object	• Only provides some views of the faces of an object • Not detailed enough for manufacturing purposes
Oblique drawing			
Orthographic drawing			
Exploded view drawing			
Assembly drawing			
Block diagram			
Circuit diagram			
Flowchart			
Wiring diagram			

Test your knowledge

What are key features of the following drawing types?

- freehand sketching
- isometric drawing
- oblique drawing
- orthographic drawing
- exploded view drawing
- assembly drawing
- block diagram
- circuit diagram
- flowchart
- wiring diagram.

3.2 Working drawings

Working drawings provide information so that products can be successfully manufactured. They contain a graphical presentation, dimensions and other important information required by the manufacturer. One of the most used working drawings in engineering is a 2D orthographic drawing. In the following section, we will look at this type of drawing in more detail, and in Unit R039, Topic area 2, you will have further opportunity to practise producing this type of drawing yourself.

Key term

Working drawing A scale drawing which serves as a guide for the manufacture of a product.

2D engineering drawings using third angle orthographic projection

A 2D orthographic engineering drawing (Figure 1.103) shows three different views (sometimes called projections) looking from the front, side and top of an object. These are shown coloured so that you can easily identify them – red (front view), blue (right side) and green (top view). This is called a third angle orthographic drawing as the side projection is usually the right-hand side of the object. In an orthographic drawing, all views are drawn on the same sheet and accurately aligned, as shown. The drawing is given within a drawing border with letters and numbers so that parts of the drawing can be easily identified. A cone and circles symbol (area D1) shows to the reader that it is a third angle projection drawing. All dimensions on the drawing are in millimetres (mm) and the drawing is drawn to scale. In this case, the bracket is drawn to actual size with a scale of 1:1. A title block (area A1) shows key details about the drawing, including the drawing number, title, scale and material for the object. A sectional view showing the object cut along the line A-A is shown in area C1–D1. You will find more details about orthographic drawing in Unit R039, Topic area 2.

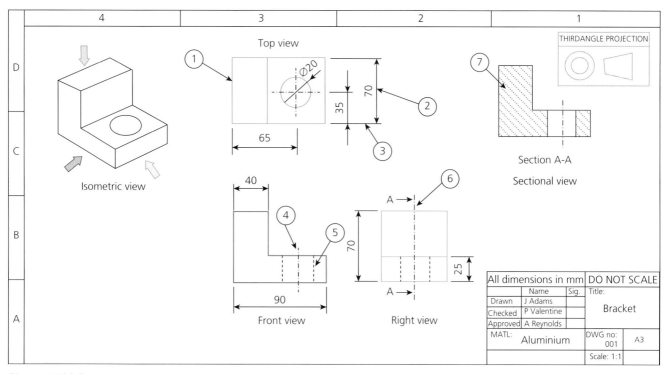

Figure 1.103 Engineering orthographic drawing

Standard drawing conventions

Engineering drawings are drawn to national and international standards using commonly understood drawing conventions. This is so that anyone picking up the drawing can understand fully what it is trying to communicate. In the UK, BS 8888 is the standard for engineering drawing. It defines very clearly what types of lines should be used, how to represent dimensions and how to indicate certain mechanical features, such as holes, chamfers and screw threads. In the next section, we will look at some of the more commonly used drawing conventions.

Line types

Drawings use different line types to communicate information. For example, a thick line is used to show the outline of objects, while a dashed line is used to communicate hidden features within the object. **Centre lines**, which show the centre of a feature, are long dashed-dotted lines. Table 1.6 shows different line types and their application. Each is numbered and refers to the lines used in the bracket drawing in Figure 1.103. See if you can identify each line type.

Key term

Centre lines Lines drawn to indicate the exact centre of a part; always drawn using a series of shorter and longer dashes (or longer dashes and dots).

Table 1.6 Drawing line types

Number	Representation	Description	Application
1	────────	Continuous wide line	Visible edges and outlines of objects to make them stand out
2, 3, 7	────────	Continuous narrow line	Used for dimension lines, extension lines, leader lines, hatching and projection lines (to help with drawing construction)
5	– – – – – –	Dashed narrow line	Used to show hidden detail in a drawing
4	—·—·—·—·—	Long dashed-dotted narrow line	Used to show the centre of a feature on a drawing, like a hole
6	▬·▬·▬·▬·	Long dashed-dotted wide line	Shows the position of a cutting plane for a sectional view

Key terms

Projection lines Lines used to extend existing lines on a drawing and used to help create new geometry.

Linear measurement Measurement indicated in a straight line.

Projection lines

Projection lines are narrow lines used to extend existing lines in a 2D drawing and are useful to help create new geometry. Sometimes they are called construction lines. Figure 1.104 shows projection lines being used to create a 2D orthographic drawing. A line is first drawn at 45° from the front view of the object (the bracket from Figure 1.103) and projection lines are drawn horizontally and vertically to help construct features on the other two views (top view and right view).

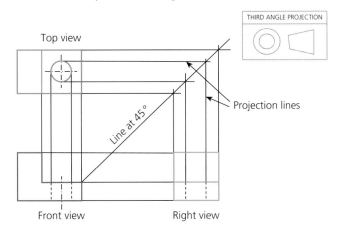

THIRD ANGLE PROJECTION

Top view

Line at 45°

Projection lines

Front view   Right view

Figure 1.104 Projection lines

Activity

Copy and complete the missing view (projection) in the following orthographic drawing. You could use projection lines to help you complete the view.

Top view

Front view   Right view

Linear measurements

Linear measurements (measurements in a straight line) are added to key features on a component on the drawing using a straight line with arrows at each end, as shown in Figure 1.105. The length (in mm) is added above the line – in this case, 90 mm. Extension lines show exactly which feature of the object the dimension relates to. The extension lines do not touch the object and a small gap is left, as shown.

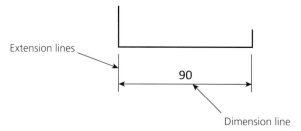

Figure 1.105 Linear measurement

Leader lines

A **leader line** is a line that points to a significant feature on a drawing. For example, it could indicate the dimension of a diameter or radius, or point out any other key feature. It is terminated with an arrow when pointing to an edge and a dot when pointing to a face. Leader lines are drawn at an angle, not horizontal or vertical. Some examples of leader lines are shown in Figure 1.106.

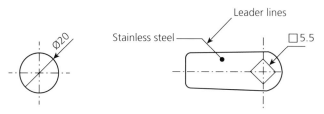

Figure 1.106 Leader lines

Key terms

Leader line A line that points to a significant feature on a drawing.

Surface finish Nature of a surface, defined in terms of its roughness, lay (surface pattern) and waviness (irregularities in the surface).

Diameter and radius

The diameter or radius of a feature on a drawing is indicated as shown in Figure 1.107. Note the use of ∅ to show diameter and R for radius. Note also the use of leader lines pointing to the diameter or radius. The dimensions are in mm, with ∅ showing that the hole has a diameter of 20 mm.

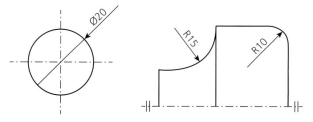

Figure 1.107 Diameter and radius

Tolerances

A tolerance shows the permitted difference in a dimension that is allowed (there is more on tolerances in Unit R039, Topic area 2). The tolerance is indicated by additional numbers next to the dimension, as shown in Figure 1.108. In this example, the dimension is allowed to be 30 mm + 1 mm, or 30 mm – 2 mm. This means that it can be in the range 28 mm to 31 mm when the component is manufactured and is still considered accurate.

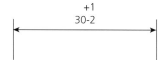

Figure 1.108 Tolerances

Surface finishes

In engineering, **surface finish** (sometimes called surface roughness) is a measure of the texture of a surface. Different engineering processes, including machining, will produce different surface finishes – sometimes rough and sometimes smooth. Surface finish is defined as the vertical deviations of a measured surface from its ideal form and is indicated on an engineering drawing using the tick-like symbol shown in Figure 1.109. Letters and numbers are placed alongside the symbol, as shown in the drawing of the object, and these show how rough or smooth the feature on the object needs to be.

Any manufacturing process permitted

Material shall be removed

Material shall not be removed

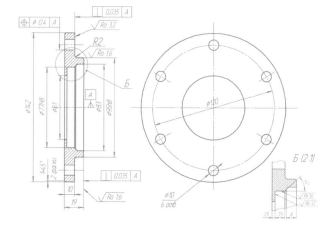

Figure 1.109 Surface finish symbols

Common abbreviations

Table 1.7 shows some commonly used abbreviations in engineering drawings, examples of how they are shown and their application. We have already come across some of these, like the diameter symbol ∅, and the use of DWG and MATL in the drawing title block.

Representation of mechanical features on drawings

Table 1.8 shows how a selection of other mechanical features are commonly represented on engineering drawings.

Table 1.7 Common drawing abbreviations

Term	Abbreviations	Example	Application
Across flats	AF	Across flats	Width across flats is the distance between two parallel surfaces on the head of a screw or bolt, or a nut as shown.
Centre line (or centreline)	CL C/L	CL	A centre line is used to show the centre of a feature, such as the centre of the hole shown in the example.
Diameter	D ∅	∅20	The diameter of a feature, such as a hole, is often represented by the ∅ symbol to indicate the dimension, as shown in the example.
Square	□	□5.5	Used to indicate the dimensions of a square feature. This saves dimensioning all sides of the square. In the example, the square cut out is 5.5 mm x 5.5 mm.
Drawing	DWG DRG	All dimensions in mm — DO NOT SCALE. Name/Sig. Drawn: J Adams. Checked: P Valentine. Approved: A Reynolds. Title: Bracket. MATL: Aluminium. DWG no: 001. A3. Scale: 1:1	Shorthand for 'drawing'. Can be used anywhere on a drawing, including in a filename extension (such as bracket.dwg). The example shows it being used in the title block.
Material	MATL Matl		Shorthand for 'material'. Can be used anywhere on a drawing to indicate material to be used. The example shows it being used in the title block – MATL: Aluminium.

Table 1.8 Further mechanical drawing features

Feature	Example	Application
Threads	External thread Internal thread	Internal and external screw threads, as shown, are represented by both solid and dashed lines. They show the screw thread on a screw or bolt, or an internal hole in which a screw thread is created using a thread tap.
Holes	Ø10 15	Holes are shown as a solid circle or using dashed hidden detail lines. In this example, the hole diameter is 10 mm and the depth of the hole is 15 mm.
Chamfers	10 × 45° 15 × 15	A **chamfer** is a cut-away at a corner of an object. Its dimensions are shown using the chamfer length and angle, or the width and depth of the chamfer.
Countersink	Ø5 Thru Ø10 × 120° CSK	A **countersink** is a bevel (or slope) on the rim of a hole so that a screw or bolt can be inserted flush with the surface. It is indicated on a drawing as two solid concentric circles, usually with the diameter of the hole and countersink and bevel angle. Note the use of CSK to abbreviate countersink, and THRU to indicate that the centre hole goes the whole way through the object.
Knurls	Knurl	**Knurling** is a process of machining a series of criss-cross ridges around the edge of something so that it is easier to grip (such as a control knob). It is shown on an engineering drawing as a series of criss-cross lines and is often labelled.

Key terms

Chamfer A transitional edge (or cut-away) between two faces of an object.

Countersink A conical hole cut into an object so that a bolt or screw can be sunk below the surface.

Knurling A manufacturing process, typically done using a lathe, where a pattern of straight, angled or crossed lines is rolled into the material.

Activity

Draw a third angle orthographic projection drawing of the object shown below. Don't forget to add dimensions and to use standard drawing conventions to complete your drawing.

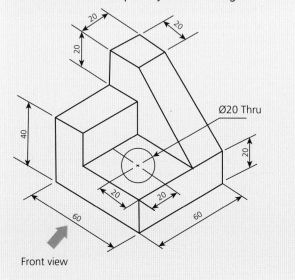

Front view

Test your knowledge

1 What three views (or projections) are shown in a third angle orthographic drawing?
2 Draw the different line types used and explain what they are.
3 How are linear dimensions, diameter and radius shown on a drawing?
4 What is a projection line and a leader line?
5 How are tolerance and surface finishes shown on a drawing?
6 What do the following abbreviations mean?
   - AF
   - CL
   - Ø
   - □
   - DWG
   - MATL
7 How are threads, holes, chamfers, countersinks and knurls shown on a drawing?

3.3 Using CAD drawing software

Advantages and limitations of using CAD drawing software compared to manual drawing techniques

Engineering drawings were traditionally produced by hand on paper attached to a drawing board using a range of drawing equipment (such as pens and pencils, rules, squares and compasses). To achieve high-quality engineering drawings that could be easily understood, the draughtsperson (person doing the drawings) would be trained to draw with meticulous attention to detail using internationally recognised standards (such as the British Standards BS:8888) for line type, thickness, image position, and so on. Following the introduction of personal computers and the development of high-quality computer-aided design (CAD) drawing software, engineering drawings can be produced to equally high standards on a computer monitor; most engineering drawings are now produced using this method.

There are numerous CAD software packages available and they mostly operate using similar sketch-based drawing or solid modelling techniques (further details in Unit R039, Topic area 3, page 136). A person skilled at using one type of CAD software package can often quickly learn to sketch and draw using a different CAD software package.

Table 1.9 The advantages of using CAD drawing software compared to manual drawing techniques

Cost	CAD drawing software is competitively priced and often available for use on a range of operating systems.
	CAD drawings can be produced in a range of locations without the need for a dedicated drawing environment.
	CAD drawings do not need to be printed on to paper and stored.
Time	Engineering drawings can be created and amended more quickly using CAD and drawing details can be modified by changing data inputs. For example, using CAD, line dimensions can be set, or amended, by digital input rather than manually measuring, marking and then drawing.
	Engineering drawings can quickly and easily be produced in different drawing and virtual modelling formats on templates that are within the CAD software. For example, 3D models can be produced from 2D orthographic CAD drawings.
	Virtual models can be made to appear as a specific material or colour, and backgrounds can be transformed to enhance the quality of the model.
Training	CAD drawing software training is often built into the cost of the software.
	Skills are often taught and learned in units and training can be available across several digital platforms in a range of locations, whereas traditionally, engineering draughtspersons required several years of training to be able to produce high-quality drawings.
Access	CAD drawings can be saved using several different methods, such as hard-drive, memory stick, cloud and website.
	CAD drawing files can be sent by email and accessed at different locations – anywhere where there is an internet signal.
	CAD drawings and models can be saved as computer-aided manufacturing (CAM) files or imported into CAM software for machining on a computer numerical controlled (CNC) machine tool, such as a lathe, milling machine, router and tool centre.
	CAD drawing files can be sent by email to remote CAM machines.
Quality	Most CAD software can produce drawings and models that can be presented as high-quality 3D virtual designs; full 360° access to models can be available in all directions.
	When combined with CAM and CNC machines, CAD-produced drawings and models can be used in the manufacture of multiple copies of identical products to a very high level of accuracy.

Table 1.10 The limitations of using CAD drawing software compared to manual drawing techniques

Cost	Computer hardware requires expensive high processing power to be able to run high-quality CAD software.
	Computers running CAD software have a limited lifespan.
Training	Changing from one CAD software package to another may require some re-training, whereas once trained, a draughtsperson has life-long abilities.
Time	CAD software is not always suitable for quick sketching and drawing; CAD drawing format, templates and tools must be selected prior to drawing.
Access	Computers running CAD software can be subject to software and power failures.

Research

Use the internet to find examples of CAD drawings and modelling, CAM and CNC machining.

Test your knowledge

1 What are the advantages of using CAD drawing software compared to manual drawing techniques?
2 What is the downside of using CAD drawing software compared to manual drawing techniques?

Practice questions

1 What does the following symbol on an engineering drawing represent? [1]

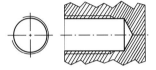

2 Which mechanical feature, which appears on an engineering drawing, is shown in the diagram? [1]

3 Name two types of view that can be included on an engineering drawing. [2]

4 State two pieces of information that may be added to the title block on an engineering drawing. [2]

5 Explain why exploded views are used in engineering drawings. [3]

Topic area 4 Evaluating design ideas

Getting started

Choose two similar consumer products and suggest which is the better product (for example, types of hairdryers, games consoles/controllers, Bluetooth speakers or mobile phones):

- Describe the design features and details that, in your opinion, make one a better product than the other. Do not comment on the brand.
- List the functions of both products.
- Comment on the purpose of each product and how well it fulfils this purpose.

4.1 Methods of evaluating design ideas

Evaluation is one of the most important processes in product design, development and manufacture. To be successful, businesses have to review their processes and outcomes to ensure they are as effective and accurate as they can be. An engineer should also review the quality of the processes they have used and the quality of outcome, to identify if and where improvements could have been made. Some evaluations will be subjective, which means they are based on personal opinions and possibly biased views, while others will be objective, which means they are factual and criteria-based.

Subjective evaluation is based on how the reviewer sees and understands something, and whether they personally like it or dislike it. The actual quality of the process or item being reviewed may not be obvious, or indeed important. Retailers depend on consumers making subjective evaluations when they are looking to buy items such as clothes, household goods and furnishings. This is one reason why there are consumer protection regulations and safeguards – to ensure quality standards are maintained.

Key term

Subjective evaluation Appraisal based on personal views, which may include bias.

Consumer associations and magazines such as Which?® aim to take some of the chance element out of buying products, by reviewing them against fixed criteria and comparing them against the competition. The criteria chosen for **objective evaluation** will be similar to a manufacturer's design specification criteria.

Key terms

Objective evaluation Appraisal that is based on fact, is reliable and could be repeated if performed by another person.

Summative evaluation Appraisal at the end of a series of processes.

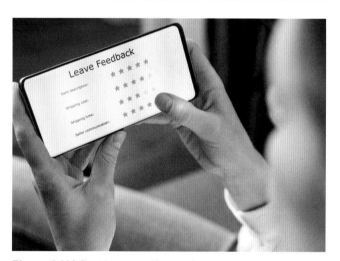

Figure 1.110 Posting an online review

Production of models

When evaluating a completed model, you should remember the processes involved in making the model – that is, research, planning, materials and tools management – because the quality of the completed product will often reflect the quality of the combined processes. A **summative evaluation** of planning will look back at the flowcharts, Gantt charts and planning for making documents to identify where there may have been miscalculations or errors that could be corrected if the engineer started the whole process again. Where a flowchart is found to have errors, they

can be overwritten, or cut and pasted with the corrected flow paths inserted. A Gantt chart can be marked up with actual timings alongside the planned timings, to help evaluate how realistic they were. See Unit R040 for further detail.

Qualitative comparison with the design brief and specification

When comparing the outcome against the design brief and specification criteria, the engineer is trying to provide an objective evaluation that would be the same if performed by another person. The evaluation comments should be factual, relevant and fully explained (justified), with sufficient detail to support the judgement.

For example, a specification criterion for a bicycle light states that it must provide appropriate lighting from high-output LEDs. Therefore, the bicycle light should be evaluated to see if it does this. In order to provide factual, relevant comments, the bicycle light will be inspected and tested to answer the following questions:

- Have LEDs been used in the construction?
- Do LEDs provide appropriate lighting?

These are relevant and easy to answer with facts. Justification can be provided by explaining how the results were achieved – that is, explaining how the bicycle light was inspected and tested.

The design brief and specification contain many of the points that can be evaluated, particularly the critical (must) criteria, and ACCESS FM can once again be used to help organise the evaluation.

Table 1.11 addresses each ACCESS FM (and manufacturing) point in turn, with key questions followed by some supplementary questions. The aim of the evaluation is to identify good qualities and to find areas that could potentially be improved. The supplementary questions will help you to interrogate beyond the key question; as the evaluation develops, so may the number of key and supplementary questions.

Table 1.11 Using ACCESS FM and Manufacturing for evaluation

Issues for evaluation	Questions to ask
Aesthetics	Is the prototype model visually appealing?
	How do shape, colour, texture and tone add to or detract from the design?
Cost	Has the prototype model been made within the agreed budget?
	What was the target cost?
	Were any unplanned costs incurred?
	What wastages were there, and how could waste have been eliminated?
Customer	Have customer needs and wants been met?
	Is the prototype model appropriate considering:
	• age range • gender (if relevant) • lifestyle • geography • hobbies or interests?
Environment	Does the prototype model perform as intended in its working environment? (How?/Why not?)
Safety	Is the prototype model safe for use?
	Does the model comply with appropriate standards or regulations?
	Do safety features function as intended?
	Will the model be reliable over time?
	Are its materials non-toxic, resistant to corrosion, non-flammable and resistant to the stresses that may be applied?
Size	Has the prototype model been made within size limitations?
	What were the agreed tolerances and how close to tolerance is the model?
	If a tolerance has not been met, how does it affect performance of the model?
	Did any tolerances need to be changed?
	How easy is it for the average person to interact with the prototype (use it, operate it, open/close it, lift it, place it down)? Consider average measurements (such as height, hand or foot size) for both males and females and how comfortable the prototype is to use.
Function	Does the prototype model do what it is designed to do?
	How well does the model perform considering issues such as:
	• environment of use • weight • strength • serviceability • maintenance • cost • resistance to corrosion • water resistance • operating temperature • number of times it can operate before failure • durability • flammability? How successful are the features and details (such as switches, lights, dials, inputs/outputs)?

Issues for evaluation	Questions to ask
Materials	How appropriate are the materials that were selected for the prototype model?
	How well do the materials perform considering issues such as:
	• environment of use
	• weight
	• strength
	• serviceability
	• maintenance
	• cost
	• resistance to corrosion
	• water resistance
	• operating temperature
	• number of times it can operate before failure
	• durability
	• flammability?
Manufacturing	How successful were the prototype modelling processes?
	Has the prototype model been produced accurately to engineering drawing tolerances?
	Could other manufacturing processes have produced a better quality of prototype?
	What other manufacturing techniques could be used to produce the prototype (such as computer-controlled, rapid-prototyping, automated processes)?
	Were any computer-controlled processes used in prototype modelling?

In order to answer some of the evaluation questions, you may need to develop tests (such as for function, safety, measurement) or to seek the views of others (for example, through a survey or questionnaire).

Ranking matrices and quality function deployment (QFD)

When evaluating, it can be useful to look back at research tools that were used during the identify and design phases and use them to evaluate the completed prototype model.

The designer will have used **ranking matrices** to compare existing products to identify and rank their features, characteristics, strengths and weaknesses. The newly completed prototype model can be evaluated and ranked against the same existing products using the same comparison terms (such as ACCESS FM) and tools (such as matrix and bar chart). This will show how the new prototype compares against existing products. Clearly, if the prototype model is a concept model (that is, it looks like the final product but without all the functions), some of the functional comparisons may not be applicable.

Before making the prototype model, the designer may have identified customer needs and wants through a quality function deployment (QFD) study. If the prototype model is evaluated against the same House of Quality matrix (see page 159), this could indicate how and where customer priorities have been achieved.

Evaluation against initial ranking matrices and the House of Quality matrix may point to suggestions for further improvements to the prototype model, so it is essential that the evaluations are fully explained so that suggestions can be justified.

Further information on ranking matrices and QFD can be found in Unit R040 (page 158).

Key term

Ranking matrix A table with numbers assigned to rate product features and used to compare products.

Stretch activity

Many commercially produced products undergo updates and model refinements to match cultural and fashion trends. Identify two products that have been updated (sometimes called upgraded) and comment on these updates.

Test your knowledge

1 Why can subjective judgements be unreliable when evaluating a prototype?
2 Why is it useful to review planning documents after a prototype has been manufactured?
3 What are the key areas for evaluation following manufacture of a prototype?
4 Why should evaluation comments be justified?
5 What evaluation test can be applied to a completed prototype?

4.2 Modelling methods

As we have seen, virtual and/or physical models of a product are often produced in the optimise and validate phases of the iterative design process. They can also be produced in many of the other design strategies, such as linear design, inclusive design and ergonomic design. Modelling allows the designer to find out ways in which the design can be improved and can be used to get feedback from the client and potential customers to inform design modifications.

There are many different methods for creating both virtual and physical models and prototypes. For physical models, the methods and materials selected depend on what the model is trying to demonstrate and the information the designer wants to obtain. Each method has its own advantages and disadvantages. Virtual and physical modelling are covered practically in Unit R040, including the planning, processes and safe use of tools needed when making physical models. Here is a summary of the different modelling methods.

Virtual (3D CAD)

Virtual modelling uses computer-aided design (CAD) software so designers can create virtual prototypes to show concept designs of products to clients and customers for their feedback. It can be used to show realistic images of the product, animation and how products are assembled. Advanced software even allows design engineers to perform virtual testing to show how parts move in relation to each other, and to simulate things like mechanical forces or flow of air across the virtual model.

Virtual and augmented reality is now a popular way for visualising and testing design ideas, as shown in Figure 1.111.

Figure 1.111 Augmented reality virtual model of a car

Virtual modelling has many advantages, including:

- It is much quicker and requires fewer practical workshop skills than producing physical models.
- It takes less time to create, test and modify a wide range of different design ideas.
- It costs less to explore designs compared to creating physical models.

Disadvantages include:

- Designers need training and skills to be able to use virtual modelling software correctly.
- Software often needs to operate on powerful computers, which cost more.
- There is no physical product model to hold, feel, measure and test.

Card

One of the simplest and quickest ways to produce physical models is to use **sheet** materials such as card. Card is inexpensive and available in a range of types, thicknesses and colours. It is also easy to work with and to shape into complex models using simple tools, equipment and processes. It allows physical models (often scale models) to be created quickly. While the models are not made of the materials that will be used in the final product and will often not function, they do allow the designer, client and customers to visualise and feel the product. They allow the proportions and scale of the product to be explored. A simple card model is shown in Figure 1.112.

Figure 1.112 A card model of a car

Advantages of modelling using sheet material such as card include:

- Materials are inexpensive and available in a range of types and colours.
- It requires only simple tools and processes (such as scissors, craft knife, guillotine, adhesives/glue, mechanical fixings such as butterfly clips).
- Models are relatively quick to produce.

Key terms

Sheet A flat material up to approximately 10 mm in thickness.

Block A rigid piece of material that is supplied with relatively flat surfaces.

Disadvantages include:

- Models are often not functional so it is not possible to test them.
- Models are not made of the production materials (such as metals and woods), so they do not fully look and feel like the final product.
- This method is only useful for relatively simple models, as not all details shown.

Block

Block materials, such as expanded polystyrene modelling foam, wood or metal, can also be used for producing models. While foam models, like the one shown in Figure 1.113, are relatively easy to produce using simple tools, models made from wood or metal require more skill and equipment to manufacture. Models made using block materials often have more 3D detail than simple card models and can be painted and finished to look and feel more like the final product. If they are made from the actual materials that the designer thinks the final product will be made from, then they can be truly realistic and might even be made to function.

Figure 1.113 Blue modelling foam model

Advantages of using block materials to make models include:

- Realistic and detailed models can be created.
- Models can be made to look and feel like the product.
- Models can be made to function and operate.

Some disadvantages are:

- Materials like wood and metal require dedicated tools, equipment and skill to process into a model (see Unit R040 for more details on the tools and equipment required).
- Making more detailed or complex models is time-consuming and expensive.

Breadboarding

Breadboarding is a technique used to construct prototype electronic circuits. There are two types of breadboard: temporary breadboards, as shown in Figure 1.114, where solderless connections are made using interconnecting wires, and a more permanent type where electronic components are soldered to the board.

> ### Key term
>
> **Breadboarding** The construction of an electronic circuit on a board (solder or solder-free) using jumper wires to transfer voltage around the breadboard.

With solderless breadboards, electronic components and linking wires can be inserted by a push fit into a board that has lines of parallel connector strips underneath the surface. Solderable breadboards are like solderless breadboards in that they also have lines of parallel connector strips. Components and linking wires can be inserted through holes in the breadboard and soldered to form a permanent and more durable circuit board.

This method of circuit construction is often used in the initial stages of design for trying and testing ideas. Circuits can be easily modified without the need to produce time-consuming and expensive printed circuit boards (PCBs).

As with other prototyping and modelling methods, using breadboards to design, construct and test circuits has its advantages and disadvantages.

Figure 1.114 Solderless breadboard with electronic components

Advantages include:

- It is quick and simple to build and connect circuits ready for testing.
- It is easy to make circuit amendments and changes.
- This method costs less than producing PCBs.
- Breadboards can be reused many times for different circuits.

Disadvantages include:

- This method does not work well for complex circuits with many link wires.
- It is a non-permanent solution so if dropped or moved link wires could come loose.
- Fault-finding can be difficult in complex circuits with lots of link wires.
- Breadboards are physically much larger than the same circuit on a PCB.

3D printing

3D printing is an additive-manufacturing process where components and products are manufactured by building up materials such as polymers and metals. An example is shown in Figure 1.115. This is unlike traditional processes where material is subtracted from a larger piece of sheet or block material to leave the shape of the desired component or product.

While 3D printing is often used to make models and prototypes of products as part of the design process, it is becoming increasingly popular to make functional components for final production products.

Figure 1.115 3D printing

3D printing covers a whole family of different ways in which to create components through additive manufacturing, including fused deposition modelling (FDM), stereolithography (SLA) and selective laser sintering (SLS).

Advantages of 3D printing include:

- It provides quick production of prototype components and products – print on demand.
- It is easy to make design changes and print another 3D model.
- Components with shape and geometry can be created that would be impossible using traditional methods.

- It is a cost-effective method as there are fewer waste materials with additive processes.
- It is an environmentally friendly method.

Disadvantages are:

- There are limited materials currently available for 3D printing (some plastics and metals, but not all).
- Sometimes parts are not fully accurate, or have poorer quality of finish, than traditional manufacturing (depending on the 3D printing process used).
- Components aften need cleaning up to remove surplus support materials once printed.
- 3D printing is often quite slow, so not ideal for large volumes of production.
- Equipment can be expensive and skills are required to set up and operate the equipment.

Research

Use the internet to find out about the following types of 3D printing and how they work:

- fused deposition modelling (FDM)
- stereolithography (SLA)
- selective laser sintering (SLS).

Activity

Draw a table like the one below and summarise the key characteristics, advantages and disadvantages of each of the different methods of modelling. One has been completed for you.

Type	Key characteristics	Advantages and disadvantages
Virtual (3D CAD)		
Card/sheet materials	• Models are made from card or other sheet materials. • Material is cut to shape and joined to other pieces to make the model. • The model will often be a smaller (scaled) version of the product.	Advantages: • Low-cost materials are available in a range of types and colours. • It only requires simple tools and processes. • Models are relatively quick to produce. Disadvantages: • Models are often non-functional. • Models are not made from production materials so they will not look and feel like the final product. • It is only useful for relatively simple models.
Block materials		
Breadboarding		
3D printing		

4.3 Methods of evaluating a design outcome

Evaluating the design outcome is part of the validate phase of the iterative design process. The designer performs testing of the virtual and/or physical prototype to ensure it meets the requirements of the client's design brief and design specification and to get useful feedback from users. They use the outcomes of testing and evaluation to make design modifications and improvements.

Methods of measuring the dimensions and functionality of the product

Measuring dimensions

One of the most fundamental checks that is made on a model or prototype product is measuring its dimensions or other parameters. The designer uses measuring instruments to check key dimensions and to check that these meet the design brief and specification. There are many different types of measuring instrument, each with its own relative benefits and disadvantages for taking measurements. Here are a few of the more commonly used instruments.

Steel rule

A steel rule is perhaps one of the most basic and useful measuring devices. Unlike the plastic rule that you will already be familiar with from your pencil case, a steel rule, like those shown in Figure 1.116, allows for more accurate measurements. Sometimes the steel rule is

called an engineer's rule. Steel rules come in many different lengths and have a precision scale on one or both sides, usually in millimetres and centimetres. Advantages of using a steel rule include it being low cost and easy to use. Key disadvantages are that they can only measure to an accuracy of 0.5 mm, are only useful for measuring small distances (up to about 1 m) and can sometimes be incorrectly read by the user. The steel rule is good for measuring external lengths, widths and depths of an object, but not so good for measuring more complex features like diameters or internal dimensions.

Figure 1.116 Steel rules

> ## Key term
>
> **Vernier caliper** A measuring device that consists of a main scale with a fixed jaw and a sliding jaw with an attached measuring scale called a vernier.

Vernier caliper

A **vernier caliper** is used to take very accurate measurements, typically up to ± 0.01 mm. It can be used to measure outside dimensions (as shown in Figure 1.117) using the external measuring jaws or internal dimensions using the internal measuring jaws, which you can also see in the image. Vernier calipers are available in many different types and although some can measure up to 1 m or more, a common maximum is 300 mm. Measurement scales are read either manually by the user or as a digital readout, which is simpler to use. Those with a manually read scale (called a vernier scale) take some skill and practice to use. Advantages of the vernier caliper include its large measurement range, that it can take accurate measurements and that it can take both external and internal measurements. Disadvantages are that it requires skill and practice to use correctly, especially those with a vernier scale, and it is a more expensive measuring instrument than a steel rule.

Figure 1.117 Vernier caliper

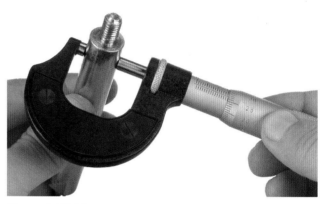

Figure 1.118 Micrometer

Micrometer

A **micrometer** is a precision measuring device. It measures objects inserted between two faces of the device, as shown in Figure 1.118, and it works well for measuring the diameter of round objects. It uses a precision screw thread which is turned to move one of the faces in and out to just touch the object being measured. The distance between the two faces is recorded on a measuring scale, which the user needs to read. Like the vernier caliper, micrometers are also available with a digital readout. The micrometer is much more accurate than both the steel rule and the vernier caliper, being able to measure as accurately as ± 0.001 mm. However, it does have a limited measurement range of typically 25 mm. Micrometers are available that can measure both external and internal dimensions. Advantages of using the micrometer include its high accuracy and, in particular, its suitability for measuring round objects. Disadvantages are its limited maximum measurement range, skill required to use and read it correctly, its cost and its robustness (it is easily damaged if dropped or used incorrectly).

Activity

Draw a table and summarise the key features and relative advantages and disadvantages of the steel rule, vernier caliper and micrometer for measuring the dimensions of a product or component.

Multimeter

So far, we have only considered using measuring instruments that record dimensions involving distance. When testing a model or prototype, there are many other parameters that the designer might wish to measure and check against the design brief or specification. If the design includes electronic circuits, this might include voltage, current or electrical resistance. A **multimeter** (Figure 1.119) is a useful piece of electronic test equipment that allows these measurements to be made. Advantages of the multimeter include them being able to measure many different circuit parameters with the same device, as well as being small, portable and relatively low cost. Disadvantages include limitations in what they can measure and display (compared to more advanced electronic test equipment), and they may not work correctly if the internal battery is low.

Figure 1.119 Multimeter being used to test a circuit

Functionality

In addition to measuring and checking dimensions to evaluate a design, the designer will want to check that the model or prototype product functions as intended, and as specified in the design brief and design specification. Testing the **functionality** of a product against how it is expected to operate and perform can be done virtually using software or with a physical prototype.

Functionality testing can include checking that the components that go to make the product fit together correctly, and that the product operates as expected. For a cordless vacuum cleaner, this could include checking that the brushes and waste container fit correctly to the body, and that the vacuum picks up dust and fluff as expected with a minimum battery life before needing to be recharged.

Functionality testing can be done using virtual CAD modelling, where the design is simulated and functionally tested within software. You will learn more about this in Unit R040. This has the advantage that the design can be easily modified without the need to make time-consuming and expensive physical prototypes. Only once a virtual functional design has been agreed is a

physical prototype made. However, functionality testing a physical prototype has the advantage that the true dimensions and operating performance of the product can be checked in real-world conditions.

> ### Research 🔍
> Use the internet to find out more about the functionality testing of product prototypes using virtual CAD modelling and by making physical models. Produce a short presentation of the relative advantages and disadvantages of both methods.

Quantitative comparison with the design brief and specification

Quantitative data is information that is based on numbers, which can often be measured. On the other hand, **qualitative data** is information based on descriptions or observations, which cannot be measured. The designer will regularly check that the product being designed and any models of prototypes created meet the requirements set out in the design brief and design specification. This will include using quantitative data obtained from taking measurements and other checks and comparing them with the requirements for the design.

> ### Key terms
> **Functionality** The purpose for which something is designed or expected to fulfil.
>
> **Quantitative data** Data based on numbers and quantities, which can be counted or measured.
>
> **Qualitative data** Data based on descriptions and observations, which cannot be counted or measured.

An example prototype desk lamp is shown in Figure 1.120, with the engineering design specification for the lamp in Table 1.12. The designer can make a quantitative comparison of the prototype lamp against the design specification by checking and comparing all the items that can be measured or checked.

Items that can be measured include:

- target and selling costs
- overall height, width and shade diameter
- weight of the lamp
- wattage of the LED bulb
- thickness of the steel material.

Figure 1.120 A desk lamp

Anything that can be measured or checked, without any confusion or disagreement, can be considered quantitative data. This can be compared against the requirements of the design brief and design specification to check that the product meets these requirements.

Of course, there will be items in the design brief and design specification that are not so easy to compare directly against the product model or prototype. These are things that rely on the opinions of clients, designers, users or customers and are called qualitative data. For the desk lamp, there might be a wide range of opinions on who a 'traditional home office customer' is, whether the design of lamp has a 'classy look and feel' or if the 'black, glossy, smooth' finish is as required.

Table 1.12 Engineering design specification for a desk lamp

Aesthetics	Black, glossy, smooth metal, traditional Anglepoise lamp look and style
Cost	Target selling cost £50; manufacturing cost £25
Customer	Aimed at traditional home office customers; classy look and feel
Environment	Many metal parts easily recyclable; manufactured in the UK from recycled materials; replaceable energy-efficient bulb
Size	Overall height 480 mm, width 300 mm, shade diameter 90 mm, weight 0.5 kg, 5-watt LED bulb
Safety	Meets all safety standards, CE marked, WEEE marked
Function	Lamp can be switched on and off with a switch; arms and mechanism can move lamp to many different positions
Material (and Manufacturing)	Painted pressed steel, 1.5 mm thick; PVC insulated copper cables and connectors; standard fixings and components used

While both types of data have their advantages and disadvantages, quantitative data is much easier to use when comparing a design with its design brief and specification.

Activity

In pairs, consider the following requirements in a design brief for a pair of wireless ear buds and discuss which are quantitative and which are qualitative:

- dimensions (each) 17 x 18 x 41 mm
- smooth white appearance
- CE marked
- comfortable to wear and use
- long battery life
- weight (each) 4 g.

User testing

User testing (sometimes called usability testing) is a stage in the design process where a product is given to real users to test. It allows the designer to evaluate their product in real-world situations and to create products which are user-centred (we came across this earlier on when we looked at different design strategies; see page 4). It allows the designer to examine and analyse what users think and how they behave when interacting with the product.

> ### Key term
>
> **User testing** Evaluating a product by testing it with representative users; sometimes called usability testing.

Sometimes a prototype product will be given to an individual user to try out or it might be given to a group of users (a focus group) for their feedback. The users need to be representative of the type of users that are likely to use the product and include a cross-section of users to ensure a wide range of views and opinions.

While users can be shown virtual prototypes of a product, it is more common when user testing for them to be given physical prototype products that look, feel and operate as detailed in the design brief and specification. Feedback from observing users interacting with the product helps the designer to decide which design modifications and improvements to make.

User testing needs prototypes to be made, which means that it is often time-consuming and expensive if the prototypes are truly like the real product. It also takes time to recruit users and to observe them. If users find serious issues in the design following user testing, it could mean significant design changes are required. This is again time-consuming and costly, and it could delay the final product. User testing is best done as early as possible in the design process.

> ### Activity
>
> Working in pairs, discuss the advantages and disadvantages of user testing, both to the designer and to the final product design.

Reasons for modifications and improvements

Designers use the outcomes of product evaluation to identify modifications and improvements to the design of a product. All modifications and improvements might not be included in the final design if they are too difficult, expensive or time-consuming to make. Those relating to safety must always be made. Table 1.13 shows some areas where potential modifications and improvements could be made. You might be able to think of others.

Table 1.13 Reasons for modifications and improvements to a design

Area	Reasons for modification/improvement
Aesthetics	• Make design look and feel more attractive to customers. • Make design more distinctive and stand out from other designs. • Make design easier to use (such as colours and style of buttons and controls).
Ergonomics	• Ensure design fits users better. • Make sure design is comfortable to use.
Features and functions	• Improve existing features and functions to make them work better. • Add features and functions that users feel could be useful. • Remove features and functions that users don't think they will need.
Safety	• Make sure that the design is safe to use (this is an important reason for making modifications and improvements).
Product quality	• Improve the quality of the product to make it more attractive to customers.
Sustainability	• Ensure product design is sustainable (increasingly seen as important by customers).
Materials and manufacturing	• Select different materials or manufacturing processes to improve product design and manufacture.

Making modifications and improvements to a design will also impact on how much it costs to manufacture. This affects the price the customer needs to pay for the product and so there is always a fine balance between what modifications and improvements are made against what the customer expects to pay.

Activity

A computer mouse is an example of a product where over time there have been many design modifications and improvements.

Working in small groups, discuss the types of improvement that could be made to the computer mouse and the reasons why these might be made.

Test your knowledge ✔

1 How would the following be used to evaluate a model or prototype, and what are the advantages and disadvantages of each:
- steel rule
- vernier caliper
- micrometer
- multimeter?
2 What is functionality testing, and how is it used to evaluate products?
3 What is quantitative data and what is qualitative data?
4 Why is user testing often used by designers to evaluate products?
5 What are the reasons for identifying potential design improvements and modification, and why are they not always made?

Practice questions ✔

1 State what technologies could be used to produce a virtual model of a product. [1]
2 Name a 3D printing technology. [1]
3 Give two advantages of producing a prototype model using card. [2]
4 Give two methods that can be used to measure the dimensions of a product. [2]
5 Explain why a designer may need to modify the design of a new product once it has been evaluated. [4]

Synoptic links

Unit R038 helps you to develop key knowledge, skills and understanding which can be applied throughout the other units of the qualification. It supports the following areas in particular:

For Unit R039: interpreting design briefs and design specifications to produce sketches of design ideas, manual production of engineering drawings, using CAD to communicate design proposals.

For Unit R040: using research methods to investigate existing products, modelling and evaluating design ideas both virtually and physically.

Unit R039

Communicating designs

About this unit

This practical unit gives you the opportunity to learn how to communicate engineering designs through freehand sketching, formal engineering drawings and 3D computer-aided design (CAD) presentation.

You will develop skills in sketching to generate a range of initial design ideas. Then you will select ideas to develop into formal engineering drawings, using CAD and other techniques to communicate a final design proposal.

Topic areas

In this unit, you will learn about:

1   Manual production of freehand sketches
2   Manual production of engineering drawings
3   Use of computer-aided design (CAD)

How will I be assessed?

This unit is assessed through an assignment that will take place at your centre. The assignment contains a scenario and a set of tasks for you to complete. Your work will be assessed against a set of marking criteria. The time to complete the assignment is included with the assignment brief. Your teacher will give you clear guidance about the tasks required to complete the assignment and the criteria you need to meet.

Topic area 1   Manual production of freehand sketches

Getting started

Consider the basic shape of the appliances you see around you, at home, in shops or in school. You will notice that many are formed from foundation shapes, such as squares, rectangles, triangles, circles and ellipses; often they use a combination of foundation shapes. For example, many metal appliances that are square or rectangular have rounded corners; doors or drawers are square or rectangular; and controls are often circular.

Produce a quick freehand pencil sketch of an appliance and highlight foundation shapes or combinations of these shapes.

1.1 Sketches for a design idea

An initial sketch is often nothing more than a rough drawing, but it can communicate concept thoughts and ideas, without needing to be formal or accurate. Initial hand-drawn sketches can be developed further and refined to represent a design for a product.

Sketching requires very basic tools. Some engineers prefer conventional pencils (blue or graphite), while others use propelling pencils (mechanical pencils with replaceable graphite leads) or fibre-tip fine-line drawing pens. For pencil sketching, 2H, HB and 2B pencils should be sufficient. For propelling pencils and fine-line pens, line widths of 0.1, 0.3, 0.5, 0.7 mm and 0.8 mm should be adequate.

As designs are developed and refined with additional features and detail, it is often best to use a combination of both pencils and pens.

A pencil eraser is not required. At the early stages of sketching, every line on the page could be useful. Therefore, it would be a waste of time to erase lines.

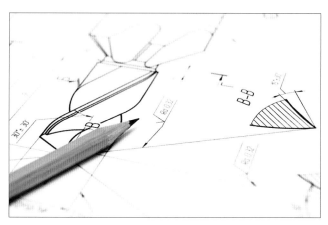

Figure 2.1 Pencil sketching

Figure 2.2 Sketching with a fine-line drawing pen

Produce a freehand sketch of a design idea

It can be daunting to pick up a pencil or pen to sketch new creative and **innovative designs**. However, sketching skills can be learned and improved with practice.

Freehand sketching in 2D

The **random line technique** is a method for producing creative and innovative initial design ideas, no matter how good a designer is at sketching. Other techniques can then be used to develop sketches and make them more formal.

In this technique you sketch multiple lines quickly in all directions across a page. These lines can be straight, circular or elliptical, or a combination of all three. By joining some of the lines together to form a boundary outline, you can produce a new creative shape. Thickening the boundary line can help to define the new shape. At this stage, the shape produced simply points towards a potential design. Within the random lines there may be many, very different potential designs.

Key terms

Innovative designs New groundbreaking, inventive designs.

Random line technique Multiple random lines are sketched in different directions across the page. Some of these lines are joined to form an outline shape for a design.

Crating A technique used to provide a framework for sketches and drawings.

Figure 2.3 Random line sketches

Activity

Divide a portrait sheet of A4 plain paper into three sections.

- In the top third of the paper, create a random pattern of straight lines.
- In the middle third of the paper, create a random pattern of circular and elliptical lines.
- In the bottom third of the paper, create a random pattern of straight, circular and elliptical lines.

Join random lines within the patterns to produce creative shapes for a multi-product charger station (for example, to charge a mobile phone, watch, ear buds and games controller). Try to be innovative with shape and form.

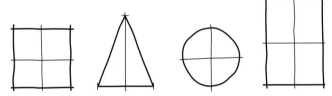

Figure 2.4 2D freehand sketches of shapes

The two-dimensional (2D) shapes shown in Figure 2.4 are the foundation of creative design: the 2D outlines of most products can be developed from these shapes. As you can see in random line sketches, the general form of most designs is rectangular, triangular or circular, or a combination of these shapes. A starting point to develop a design could be to sketch one or more foundation shapes.

- To help with proportion and alignment, it is good practice to sketch some fine **crating** lines first. These are fine construction lines that provide a framework but are barely visible on the page.
- Start with a centre line and lines to indicate height and width. Then add a few lines equally spaced within the crate.
- Don't be too concerned if lines are not as straight or as curved as you would like – at this stage, it doesn't matter.

- When sketching straight lines freehand, start by sketching part of the line and then extend it, overlapping the initial line.

- For curved lines, hold your pencil as normal and lay the base of your hand, below your little finger, on the page. Keeping the base of your hand on the page, move the pencil in an arc shape to sketch a curved pencil line on the page. You can produce a circle by turning the page to draw overlapping curved lines. It may be inaccurate but this does not matter in a sketch.

You can combine simple 2D shapes to form new creative **compound shapes**, as shown in Figure 2.5. Rectangles, triangles, semi-circles and arcs have been joined together to form designs. You may wish to add features such as handles, grips, caps, clamps, dials, switches, screens, lights and speakers. These details can help to characterise a potential product.

Figure 2.5 2D thumbnail sketches using compound shapes

These initial 2D sketches will be **thumbnail sketches** – small illustrations around 30–40 mm in height or width that are produced quickly.

These thumbnail sketches will include possible product designs. You can use **line enhancement** techniques, which involve changing line thickness or weight, to highlight these designs.

When using the **thick and thin lines technique** for 2D sketches, make the outer boundary lines

of a design thicker and bolder – approximately twice the thickness of a standard fine line. Use fine lines to show detail on a design, with slightly thicker lines to show surface details of particular interest. Where boundary edges meet another surface, such as a table or floor, use a very thick boundary line to emphasise surface contact. Use sharp B pencils or fine-line pens to produce thick and thin lines.

Weight of line is a technique used to add line thickness, strength, boldness or darkness. It can add drama to a design, particularly where edges curve or change direction, and where shadows may be formed.

There are no strict rules about line weight when illustrating detail – the main thing is how the sketch looks on the page. Notice how the enhanced sketches in Figure 2.6 stand out; the other initial sketches appear to fade into the background. You can make lines stand out more using fine-line pens, soft B pencils or marker pens.

Key terms

Compound shapes Shapes formed by the combination of two or more simple shapes such as squares, rectangles, triangles or semi-circles.

Thumbnail sketches Small inaccurate sketches of initial ideas for a design.

Line enhancement Increasing the thickness and boldness of object lines to highlight boundary edges.

Thick and thin lines technique A sketching technique in which outer boundary edges are sketched as thick bold lines and detail on the design is sketched using thin fine lines.

Weight of line How light, dark, bold or heavy a line is.

Figure 2.6 Thick and thin lines used to enhance 2D freehand thumbnail sketches

Activity

Using 2D crating techniques, produce several 2D thumbnail sketches of workshop tools (for example, bench hook, engineer's square, hacksaw, wood plane, screwdriver, G cramp).

Photocopy the sketches and save a copy for a later activity.

Add line enhancement (such as thick and thin lines, weight of line) to at least two of the sketches.

Freehand sketching in 3D

Once you have identified potential designs for a product from 2D sketching, the next stage is to develop sketches in three dimensions (3D). This adds depth to the designs.

To do this, you can use techniques including oblique, isometric and perspective sketching. As with 2D sketches, crating can provide a framework for each design. Start by sketching each design using fine construction lines. The basic rules for oblique and isometric sketching are the same as for formal drawing using these techniques. All crate lines that project away at an angle must remain at that angle, so the crate is uniformly shaped. As sketches are developed in 3D, they tend to become larger with more accurately defined detail.

Oblique sketching

When developing a sketch using the oblique technique, start by sketching a 2D crate. Then add lines of sight that project away from the crate at an angle between 60° and 30°. The depth of the sketch can vary between actual depth and a scaled or foreshortened view. The aim is to produce a realistic image of the object being sketched, with depth.

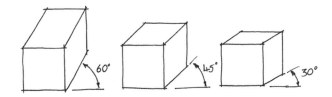

Figure 2.7 Oblique crates at 60°, 45° and 30°

Figure 2.8 shows an oblique sketch of a washing machine developed in a crate. The front view of the washing machine is sketched as a 2D image. Then a 3D crate is constructed along the lines of sight from the top and side, to add depth. When adding detail to the sketch, all vertical lines remain vertical. Arcs and circles on the 2D front of the oblique crate can be sketched in their actual shape; arcs and circles sketched along lines of sight (on the side or top of the object) should be elliptical.

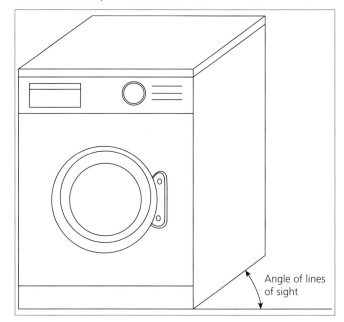

Angle of lines of sight

Figure 2.8 An initial oblique sketch of a washing machine

Thick and thin lines on 3D sketches

As with 2D sketching, you can use the thick and thin lines technique to enhance initial 3D sketches. Outer boundary lines should be thick, while detail lines may be thick or thin depending on what they are representing. There are three basic rules when using the thick and thin lines technique on 3D sketches and drawings:

- Boundary edge lines should be thick bold lines.
- Hard-edge lines should also be thick bold lines (see lines 1, 2, 3, 4 and 5 on Figure 2.9). These are lines where only one side of a line can be seen – the other side of the line is around a corner, or in a hollow or recess.
- All other lines should be standard fine lines.

3D thumbnail sketches do not need to be too accurate because the design may not be taken forward for development. Therefore they can be produced fairly quickly. You can improve 3D designs of particular interest by neatening their boundary and hard-edge thick lines using a straight edge or template.

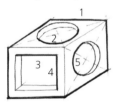

Figure 2.9 Initial 3D thumbnail sketch with thick and thin line enhancement

Key term

Vanishing point (or **viewpoint**) A point on the horizon beyond which an object can no longer be seen.

Activity

Produce a 3D oblique image of a square- or rectangular-shaped appliance (for example, a microwave or refrigerator).

- Produce the 2D crate for the front view of the object.
- Add centre lines to the 2D crate.
- Add basic detail to the front view.
- Add lines of sight.
- Add depth (the line that shows the back of the object).

Repeat the activity to produce two more 3D oblique sketches for the same appliance with lines of sight at a different angle, somewhere between 30° and 60° to the horizontal. The depth of the sketch can be foreshortened.

Photocopy the sketches and save a copy for a later activity.

Using a straight edge and templates, add neat thick and thin line enhancement to at least one of the 3D oblique sketches.

Perspective sketching

Sketching in perspective involves trying to sketch an image as the eye naturally sees it. For example, if you stand in a railway station (see Figure 2.10) and look down the length of a long train, the carriages further away will appear smaller than the carriages closer to you.

When sketching and drawing, the smallest object we can draw on a page is a dot. If we consider this dot to be the **vanishing point** on the horizon (the point beyond which we can no longer see an object) and we sketch towards this point, we can produce realistic object sketches. The nearer parts of an object are sketched proportionally larger than the parts that are further away.

Figure 2.10 A one-point perspective sketch of a train in a railway station

When you are developing a sketch in one- or two-point perspective, sketch the nearest corner of the crate first. Then sketch lines of sight away from the top and bottom of this corner, in the direction of a vanishing point (dot) on an imaginary horizon. A **one-point perspective** sketch has one vanishing point, and a **two-point perspective** sketch has two vanishing points. The vanishing point can also be called the viewpoint.

Key terms

One-point perspective A sketch with one vanishing point.

Two-point perspective A sketch with two vanishing points.

Once you have established the top and bottom lines of sight, you can develop a sketch between them without projecting every line all the way to the vanishing point(s). Curves and circular shapes appear slightly squashed when drawn in perspective, as shown in Figure 2.12.

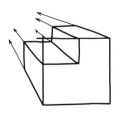

Figure 2.11 One-point and two-point perspective sketching

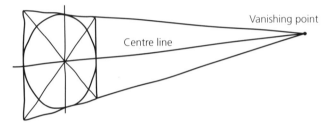

Figure 2.12 Sketching a circle in a one-point perspective

When objects are much higher than they are wide, you can use three-point or four-point perspective to produce a more realistic image. Figure 2.13 shows a three-point perspective image. Notice how the sketch also narrows towards the bottom.

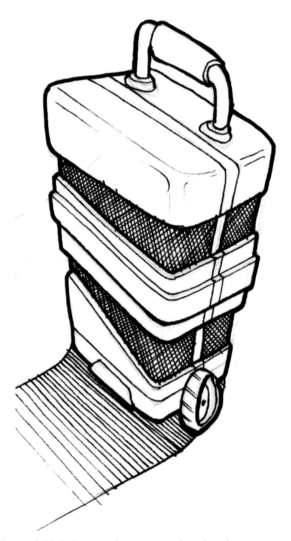

Figure 2.13 A three-point perspective sketch

Activity

Produce a two-point perspective sketch with multiple solid and hollow shapes (squares, rectangles, circles) sketched on, above and below the imaginary horizon, as shown in Figure 2.14. Use the same vanishing points for all sketches.

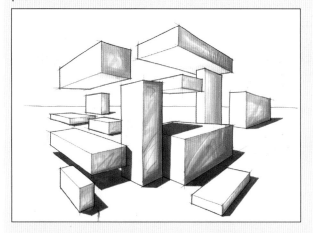

Figure 2.14 Rectangular shapes sketched in two-point perspective

Produce an isometric sketch for a design proposal

Isometric sketches are formed within a 3D isometric crate. Start a crate by sketching the nearest vertical corner. Then sketch lines of sight away from it at 30° to the horizontal (see lines 1 to 5 on Figure 2.15). Sketch vertical lines to mark the length of the sides and provide rear edge lines for the crate (lines 6 and 7). Finally, sketch lines along 30° lines of sight from the top of the rear edge lines (lines 8 and 9) to complete the crate. Sketch the isometric crate using very fine lines. Once you have formed the crate, you can develop the design for an object within it. Using this technique, all vertical lines remain vertical and all horizontal lines from the nearest corner are sketched at 30° to the horizontal. Isometric sketching uses the same scale on each side or face of the object.

Thick and thin lines can be added to enhance isometric sketches (Figure 2.15).

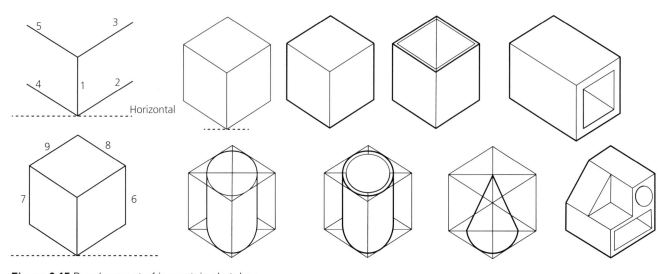

Figure 2.15 Development of isometric sketches

Use an ellipse to add circular detail on isometric sketches. You can sketch an ellipse freehand using an approximate plotting technique such as the $2/3$ method, as shown in Figure 2.16. On the longer centre line, plot two dots approximately $2/3$ of the way from the centre to each corner. Then plot a dot approximately $2/3$ of the way along each side edge.

Sketch a freehand elliptical curve to join each dot to the sides of the isometric crate. Remember, a sketch is an early illustration of a potential design so it does not need to be precise. Ellipse templates are a quick and easy alternative to plotting ellipse sketches.

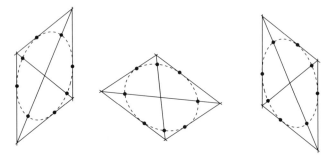

Figure 2.16 Plotting isometric curves using the $^2/_3$ method

When drawing isometric sketches on plain paper, it can be helpful to place a sheet of isometric grid paper underneath the plain paper. Small isometric grids can be used in this way when you are adding detail. You can develop complex designs by adding an isometric crate for each additional part of the sketch.

Tracing or layout paper can be very helpful when you are developing complex 3D isometric compound shapes. Sketch each component of the design to the same scale, on a different piece of tracing or layout paper. When you have sketched all the components, lay the individual sketches one on top of the other to form the full design. You can then sketch a final tracing over the top of all the underlays. Neaten the final sketch lines (thick and thin) using sketching aids such as a straight edge, French curves and ellipse templates.

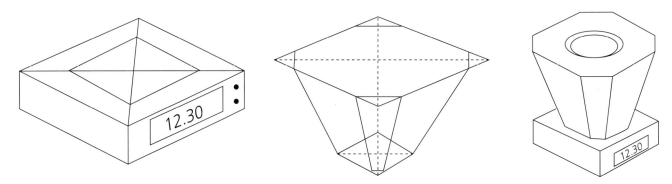

Figure 2.17 Compound shape designs developed using tracing paper

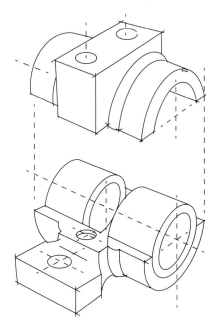

Figure 2.18 A completed 3D sketch in two-point perspective

Tracing and layout paper can also be used to lay out the position of components on an isometric **exploded diagram**. Exploded diagrams show the relationships between components of a product, or the order of their assembly. Sketch each component a short distance away from other components, so it looks as though they have been taken apart and are about to be reassembled. On products with multiple assemblies, you can show the components of each assembly close together, separate from the other assemblies.

Key term

Exploded diagram An image of a product where all the component assemblies are shown outside the product.

Figure 2.19 An exploded diagram of an LED (light emitting diode) lightbulb

Case study

Watch the YouTube video 'Product design sketching (building 3D sketches)': www.youtube.com/watch?v=JkpDCUk17K4

Discuss the techniques demonstrated in the video.

What can you learn from the video? Which techniques can you use as you develop your own design ideas?

Activity

Produce 3D isometric thumbnail sketches of:

● a cube
● a short, long rectangle
● a tall and wide rectangle with a square hollow on one face
● a cube with a circular hole through it
● a tube.

Photocopy the sketches and save a copy for a later activity.

Add thick and thin line enhancement to each sketch.

Rendering using texture, tone and shading

From a range of 2D and 3D sketches, you should be able to select a potential design for further development and **rendering**.

Key term

Rendering Application of surface decoration and detail, such as colour, shade, tone and texture.

Rendering is a process for turning line sketches into realistic design proposals by applying surface decoration and detail, for example colour, texture, tone and shading. Rendering can be added at any stage of design and development to enhance a sketch. At this stage, your line sketch design may still be quite simplistic. However, it should include some of the key elements identified in the specification criteria, such as aesthetics, ergonomics, materials and function.

Coloured, textured or plain white paper can be used for sketches alongside a wide range of rendering materials:

● Coloured pencils are often used to render sketches and they offer a good range of tone. However, low-quality pencils can be difficult to blend and erase.

● Coloured pastels can be used in many ways. You can apply them directly onto paper, scrape them to a dust and apply using cotton wool and a dry brush, or mix them with a solvent to form a paint. They also work well with other rendering techniques such as weight of line. When applied dry they can be erased easily.

● Ballpoint pens, fibre pens, roller tip pens and felt tip pens are good for general sketching and layout work. They are useful for highlighting boundaries, edging and detail, but tend to be made with permanent inks.

● Marker pens can be used to apply a wash of colour to a sketch quickly, but it can be difficult to achieve an even colour without a lot of practice. Marker pens tend to give a sense of colour rather than full colour, which can be a very powerful effect. Once marker ink is dry, other materials can be applied over the top, such as ballpoint pens, fine-line pens, felt pens and paint.

Examples of rendering techniques to communicate texture, tone and shading are shown in Figures 2.20 to 2.23.

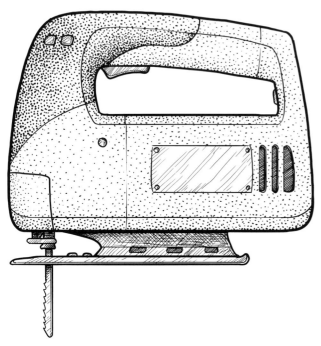

Figure 2.20 Monochrome surface rendering with line enhancement

Figure 2.21 Monochrome rendering with line enhancement

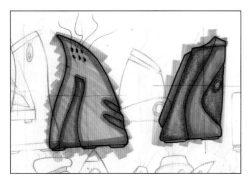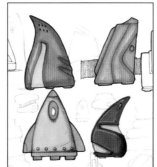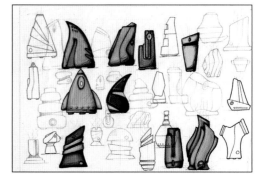

Figure 2.22 Examples of colour wash and marker rendering

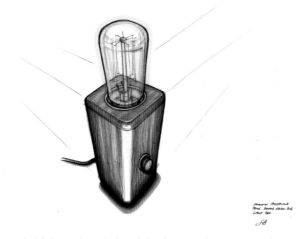

Figure 2.23 A rendered sketch in three-point perspective

Activity

Cover two sheets of A4 paper with multiple copies of the un-enhanced sketches you made for earlier activities in this unit (2D sketching and 3D oblique and isometric sketching).

Add rendering to your 2D and 3D line sketches using a range of rendering techniques and tools (coloured pencils, pastels, pens, markers) to communicate colour, texture, tone and shading.

Annotation and labelling techniques that demonstrate design ideas

While sketches can provide an illustration of what a design looks like (for example, colour, shape, detail), sketches alone are not enough to indicate the merits of a design. For a design to be successful, several, often conflicting, specification criteria should be met. It is not always obvious how a design meets specification criteria, and so they may need to be pointed out and explained. To do this, a designer will use **labels** and **annotations** that identify, describe and explain **characteristics**, features and detail.

Key terms

Labels Text added to a sketch that points out details, features and characteristics – for example, switch, LED and battery holder.

Annotations Notes or comments added to a sketch or diagram that provide explanation and give meaning.

Characteristics A noticeable or typical feature.

Annotations are notes or comments added to a sketch or diagram that provide explanation and give meaning, so that the reader can understand something about the designer's intention. For example, how a product will function and meet a specific design criterion.

Labels are quite different from annotations; they point out detail and identify components or features by name (as shown in Figure 2.24) or by number. For example, 1, 2, 3 and so on. Where numbers are used, they should match named components or features on a list.

There is no standard order when it comes to annotations because all annotations have a purpose (to explain). All annotations are important, but the design specification's essential criteria are often a good starting point. Table 2.1 identifies many of the varied design requirements a designer may need to explore. The designer should try to explain how the potential design meets a range of agenda items presented by manufactures and consumers, and each agenda item will have a range of associated requirements.

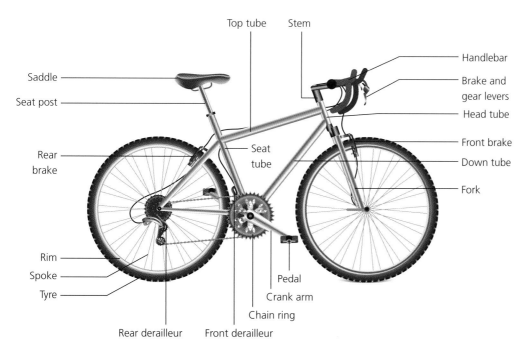

Figure 2.24 Labels being used to identify components

In Unit R040, you will learn how to analyse products using the ACCESS FM method. This same acronym can be used for annotation headings.

The **5WH method** (Who, What, Where, When, Why and How) can be useful to help develop evidence of understanding to annotations. Using 5WH questions, you can cross-examine your design proposals against specification criteria and use the answers to your questions to explain in as much detail as possible how the design proposals meet the varying demands.

For example, let's take the aesthetics (user need and product requirement issues):

- **Who** is the design aimed at and who should it appeal to (age, gender, group, ethnicity)?
- **What** are the key design features that make it appealing (shape, form, line, colour, size)?

- **Where** will it/can it be used and is it appropriate for that location (home – lounge, kitchen, bedroom, bathroom, garage, garden; workplace; car, train, bus)?
- **When** will it be used and will its visual appeal matter?
- **Why** is this design more appealing or more appropriate than others?
- **How** does the design meet conflicting needs and requirements (client, manufacturer, retailer, consumer, society)?

Key term

5WH method Questions that can be used to interrogate a design (Who, What, Where, When, Why and How).

Table 2.1 Examples of specification requirements

User needs	Product requirements	Manufacturing considerations	Regulations and safeguards	Other wider influences
Aesthetic	Function	Materials	Legislation and standards:	Market pull
Ergonomic	Performance	Scale of production		Technological push
Anthropometric	Features	Tooling	• British Standards	Sustainable design
Fashion, trends and lifestyle	Ergonomic	Finish	• Conformité Européenne (CE)	New and emerging materials
Product safety	Anthropometric	Design for manufacturing and assembly (DFMA)	• International Organization for Standardization (ISO)	New and emerging technologies
Cost	Aesthetic	Maintenance		Lifestyle
	Environment for use	Sustainability		Globalisation
	Durability	Cost		Environmental impact
	Safety			
	Life cycle			

Activity

Print an image of an electric scooter, like the one shown in Figure 2.25.

Label the image to point out specific characteristics, features and detail of the electric scooter.

Add annotations that describe the characteristics, features and detail, and describe how they could meet specification criteria.

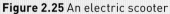

Figure 2.25 An electric scooter

Test your knowledge

1   What sketching technique can be used for producing quick creative outline shapes for products?
2   What is the purpose of sketching in crates?
3   What tools are used for rendering sketches?
4   What is the difference between a label and an annotation?
5   What criteria should annotations link back to?

Assignment practice

Marking criteria

Mark band 1: 1–4 marks	Mark band 2: 5–8 marks	Mark band 3: 9–12 marks
Produces a **limited** range of creative freehand design proposals.	Produces an **adequate** range of creative freehand design proposals.	Produces a **wide** range of creative and innovative freehand design proposals.
Limited consideration of the design specification.	**Partial** consideration of the design specification.	**Fully** considers the design specification.
Uses a **basic** range of techniques.	Uses an **adequate** range of techniques.	Uses a **comprehensive** range of techniques.

Top tips

Carefully read the assessment criteria and guidance notes for this learning outcome in the OCR specification.

Task 1

Manual production of freehand sketches

● Produce a wide range of creative and innovative 2D and 3D freehand design proposals that meet specification requirements.
● Add thick and thin line enhancement to highlight features, detail and texture.
● Use a wide range of appropriate rendering techniques to communicate colour, texture, tone and shading.
● Label design proposals to point out key characteristics, features and detail identified in the specification.
● Annotate design proposals to describe key characteristics, features and detail identified in the specification, and show understanding of client/user needs and wants.

Task 2

Manual production of freehand sketches – design development

● Through comprehensive 2D and 3D sketching and rendering, develop one design proposal further. Sketches should communicate all potential components and assembly methods.
● Label and annotate sketches to show clear evidence that specification criteria have been critically analysed. Explanations should comprehensively justify how the final design proposal meets design specification criteria.
● State where links have been drawn from skills/knowledge/understanding in other units (for example, Unit R038).

Model assignment

Hand-held games controller

- Sketch 4–6 quick freehand creative and innovative 2D designs for a hand-held games controller.
- Add line enhancement and rendering to some of the sketches.
- Label and annotate sketches to point out and explain key design characteristics, features and detail. Describe how user needs could be met.
- Select one of the 2D designs and develop it further through 2D and 3D sketches; apply appropriate rendering.
- Label and annotate the final design proposal sketches to explain key design decisions and how the design potentially meets design user needs.
- Produce an isometric sketch of the final design proposal, including key components and assembly methods.

Example candidate responses

The example candidate responses below show how a candidate has responded to an assignment about car centre consoles.

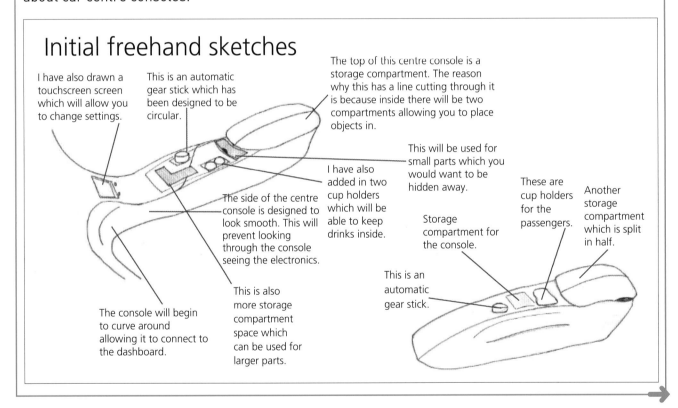

Initial freehand sketches

I have also drawn a touchscreen screen which will allow you to change settings.

This is an automatic gear stick which has been designed to be circular.

The top of this centre console is a storage compartment. The reason why this has a line cutting through it is because inside there will be two compartments allowing you to place objects in.

This will be used for small parts which you would want to be hidden away.

These are cup holders for the passengers.

Another storage compartment which is split in half.

The side of the centre console is designed to look smooth. This will prevent looking through the console seeing the electronics.

I have also added in two cup holders which will be able to keep drinks inside.

Storage compartment for the console.

The console will begin to curve around allowing it to connect to the dashboard.

This is also more storage compartment space which can be used for larger parts.

This is an automatic gear stick.

Rendered freehand sketches

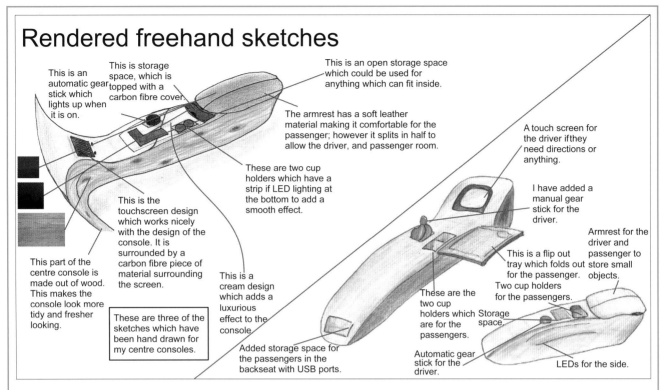

This is an automatic gear stick which lights up when it is on.

This is storage space, which is topped with a carbon fibre cover.

This is an open storage space which could be used for anything which can fit inside.

The armrest has a soft leather material making it comfortable for the passenger; however it splits in half to allow the driver, and passenger room.

A touch screen for the driver if they need directions or anything.

These are two cup holders which have a strip if LED lighting at the bottom to add a smooth effect.

This is the touchscreen design which works nicely with the design of the console. It is surrounded by a carbon fibre piece of material surrounding the screen.

I have added a manual gear stick for the driver.

Armrest for the driver and passenger to store small objects.

This part of the centre console is made out of wood. This makes the console look more tidy and fresher looking.

This is a cream design which adds a luxurious effect to the console

This is a flip out tray which folds out for the passenger.

Two cup holders for the passengers.

These are the two cup holders which are for the passengers.

Storage space

These are three of the sketches which have been hand drawn for my centre consoles.

Added storage space for the passengers in the backseat with USB ports.

Automatic gear stick for the driver.

LEDs for the side.

The candidate has enhanced 2D and 3D line sketches using weight of line techniques and rendering. They have also modified and enriched the designs.

The candidate has provided annotation to explain their thinking.

There are implicit links to knowledge and understanding from Unit R038.

Research

Research 'Producttank' videos on YouTube, including 'Product design rendering and sketching'.

Presentation techniques, Dick Powell. Orbis Publishing, 1990.

Sketching: Drawing techniques for product designers, Koos Eissen and Roselien Steur. BIS Publishers, 2013.

Sketching: The basics, Koos Eissen and Roselien Steur. BIS Publishers, 2019.

Topic area 2 Manual production of engineering drawings

For every engineered product, engineering drawings are required to describe the product in detail so that it can be successfully manufactured. In practice, drafters (those producing these drawings) take the rough sketches, specifications and calculations from designers and turn them into formal engineering drawings so that the product can be made. As a result, there are many careers on offer in engineering graphics.

Getting started

Choose an item in the room and try to produce a simple 3D drawing of it.

2.1 Drawings for a design idea

3D engineering drawings

Engineering drawings were traditionally produced by hand, using equipment such as pens and pencils, rules, squares and compasses on a drawing board and large sheets of paper. The majority are now produced using computers.

The most popular way to produce engineering drawings is to use specialist CAD software, which enables rapid development of drawings in both 2D and 3D.

However, to be able to use this software effectively, first you need to understand the basic principles of engineering drawing. Therefore, it is useful when learning these skills to have a go at producing engineering drawings by hand, before moving onto CAD.

Usually, the drafter first creates the drawing using 3D techniques, and then later presents and enhances this using 2D techniques, in order to convey enough detail for the product to be manufactured.

Several 3D drawing techniques are frequently used in engineering design.

Oblique and isometric drawings

Oblique drawing and **isometric drawing** are 3D drawing techniques that are classed as **pictorial drawing**. Pictorial drawing is used to show different views of a component or product looking from different viewpoints or angles. These types of drawing are easier to understand than just looking at the front, side or top of an object, although they do not communicate enough detail alone for it to be manufactured.

With oblique drawing, the focus is on one face of the object, while in isometric drawing the focus is on the edge of an object. Oblique drawings use an angle of 45° to the horizontal (Figure 2.26), and isometric drawings always use an angle of 30° to the horizontal (Figure 2.27).

Creating both oblique and isometric drawings requires skill and practice.

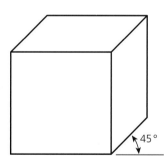

Figure 2.26 Oblique drawing

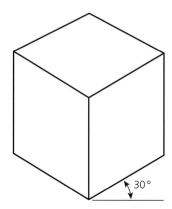

Figure 2.27 Isometric drawing

Key terms

Oblique drawing 3D pictorial drawing that focuses on the face of an object and uses an angle of 45° to the horizontal.

Isometric drawing 3D pictorial drawing that focuses on the edge on an object and uses an angle of 30° to the horizontal.

Pictorial drawing View of an object as it would be seen from a certain direction or point of view.

Distinguishing face Face or side of an object with features of most importance.

Lines of sight Parallel angled lines drawn to help construct 3D drawings.

Cavalier view Technique used in oblique drawing where all sides of the object are drawn original size.

Cabinet view Technique used in oblique drawing where the distinguishing face is drawn original size but the other sides along the lines of sight are drawn half size.

Dimensions Numerical values added to engineering drawings to communicate the sizes of key features (measurements are usually in millimetres).

Oblique drawing

There are two different types of oblique drawing – cavalier and cabinet. They both have the **distinguishing face** of the object facing the observer, with angled lines projected at 45° being used to construct the other faces. These parallel angled lines are called **lines of sight** (Figure 2.28).

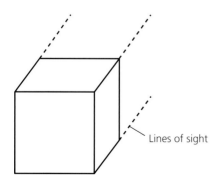

Figure 2.28 Lines of sight

In **cavalier view**, the lengths of the sides along the lines of sight are drawn at full scale (that is, at their original length). In **cabinet view**, the lengths of the sides along the lines of sight are drawn at half scale (that is, half their actual length). While both types of drawing are similar, objects often look more realistic using cabinet view.

Let's illustrate examples of each type of drawing using a simple isometric cube, as shown in Figure 2.29. Each side is 40 mm in length and there is a hole on the cube's distinguishing face. (NB It is usual for the **dimensions** of engineering drawings to be in millimetres, mm.)

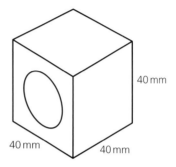

Figure 2.29 Cube of side length 40 mm with a hole on its distinguishing face

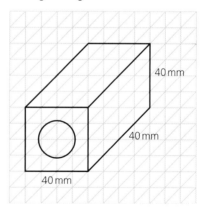

Figure 2.30 Cavalier view

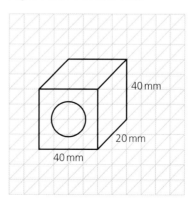

Figure 2.31 Cabinet view

Special oblique drawing paper has been used to help to produce the final drawings. Each division on this paper represents 10 mm. They could also be drawn using plain paper, a rule and a protractor to construct the lines of sight at 45°. The drawings could also be produced using CAD software.

Figure 2.30 shows the front face of the cube with the hole (the distinguishing face) drawn original size, with the other sides also drawn original size along the lines of sight.

In Figure 2.31, the distinguishing face is also drawn original size, but the other sides are drawn half original size along the lines of sight. Figure 2.31 looks more natural than Figure 2.30, which looks somewhat distorted.

While shapes with straight lines are relatively simple to produce, those with more complex shapes, such as circles and curves, are more difficult to draw (the circles become ellipses on the angled sides). Figure 2.32 shows the cube with the circular hole on the side face, which is now drawn as an ellipse.

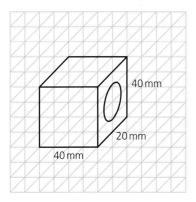

Figure 2.32 Cabinet view of a cube with a hole on its side face

Isometric drawing

With isometric drawing, the focus is on the edge of an object, with lines of sight that are always at 30° from the horizontal used to construct the other faces. Figure 2.33 shows the 40 mm cube in isometric view.

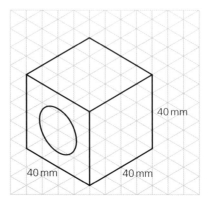

Figure 2.33 Isometric view of a cube

In Figure 2.33, special isometric paper has been used to construct the cube, which already has lines of sight at 30°. Each division on the paper represents 10 mm. Note that there is no half-scaling in isometric view, and everything is drawn original size. As before, circles and curves need to be reproduced differently on the angled faces (circles become ellipses).

Scale

It is not always possible to draw objects their actual size, as they might not conveniently fit onto the paper or screen being used to display them. In these cases, the size of the drawing needs to be reduced by a certain amount in all its dimensions. Sometimes the size of the drawing might even be increased, in order to enlarge features of interest. This information is called the **scale** of the drawing. The scale is shown as the **ratio** of the length in the drawing, then a colon (:), then the matching length on the actual object. Sometimes the scale is called the scale ratio.

Key terms

Scale Amount by which a drawing is reduced or enlarged from the size of the actual object, shown as a ratio.

Ratio Comparison of two or more numbers that indicates their sizes in relation to each other.

Figure 2.34 shows a scaled drawing of the familiar isometric cube. The side lengths are still labelled the correct size, although the first drawing is half the size of the original, full-size second drawing (the actual object). The scale is 1:2.

It is important to include details of the scale along with the drawing, so that it is clear what is being shown.

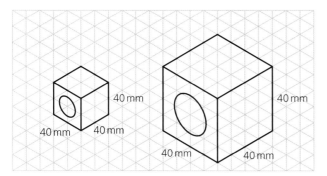

Figure 2.34 Isometric cube scale 1:2

Activity

Figure 2.35 is a simple drawing showing the front and side views of a mobile phone, along with its overall size.

Produce oblique and isometric views of the mobile phone.

You can make up your own dimensions to construct the other features, such as the display, keypad and buttons. Consider whether you need to scale the drawings.

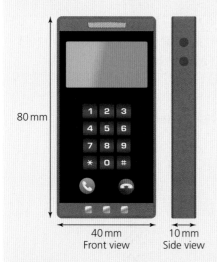

Figure 2.35 Simple line drawing of a mobile phone

Assembly drawings

Figure 2.36 shows an **assembly drawing** for a battery-powered reading light assembly, which clips onto a book. The light has many separate parts and the drawing shows how these are assembled to make the final product. Assembly drawings can include lists of parts, reference numbers, **assembly instructions** and many more details to communicate how to fit the parts together.

Assembly drawings show the parts in the correct positions and how they are assembled (or joined) to each other. The drawing in Figure 2.36 has been produced using CAD software, although it could have been hand drawn. The CAD software allows the 3D drawing to be rotated and viewed from different angles; it is extremely useful to be able to visualise the product at the design stage before manufacture.

Assembly drawings can be produced in both 3D and 2D. Figure 2.36 includes a 2D assembly drawing and a 3D isometric view. Each main part has been labelled with a number and identified in a table. Assembly drawings do not often provide sufficient detail alone to be able to manufacture each separate component. This is better explained using a series of 2D engineering drawings of each component.

Key terms

Assembly drawing Drawing showing how separate components fit together.

Assembly instructions Instructions included with a product with drawings and text on how to assemble the product.

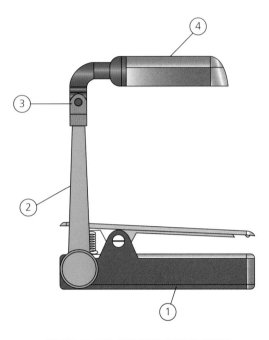

Item no.	Part name	Qty
1	Base	1
2	Arm	1
3	Swivel	1
4	Shade	1

Figure 2.36 Assembly drawings for a reading light

Exploded views

Another form of 3D drawing is an exploded view. Figure 2.37 is an exploded view of the battery-powered reading light, which shows the component parts that make up the product separated and moved outwards.

This drawing technique shows the relationship of components to each other and also the order in which they are assembled to make the final product. In Figure 2.37, each component has been labelled with a number and a table is used to identify the corresponding part name and quantity required. Note that additional components not shown in the basic assembly drawing in Figure 2.36 are now clearly visible, such as the batteries and the bulb.

Exploded views can be produced in both 3D and 2D and show more clearly the order of assembly than an assembly drawing. As with assembly drawings, an exploded view still does not provide sufficient detail to enable each separate component to be manufactured – this again requires separate detailed 2D engineering drawings for each component.

Materials, parts lists, labels and annotations

Engineering drawings can communicate many details to enable the successful and accurate manufacture and assembly of components and products. This includes the parts and materials from which the product is made. Labels and annotations can be added to communicate any further details or special instructions.

Figure 2.38 shows an exploded isometric drawing of a battery-powered torch with a swivel head. Each component on the drawing is labelled with a number and named in a **parts list**, along with the materials from which it is manufactured. This parts list is displayed in a table. The quantity of each component is also listed in the table (for example, the torch has two batteries and two locking pins). Note that part number 1 – the torch body – is made up of several components; this is called a sub-assembly. The parts list is also sometimes called a **bill of materials**.

Key terms

Parts list List of all the individual components required to make a working product, which usually includes component names, quantities, materials, costs and suppliers.

Bill of materials Another name for a parts list.

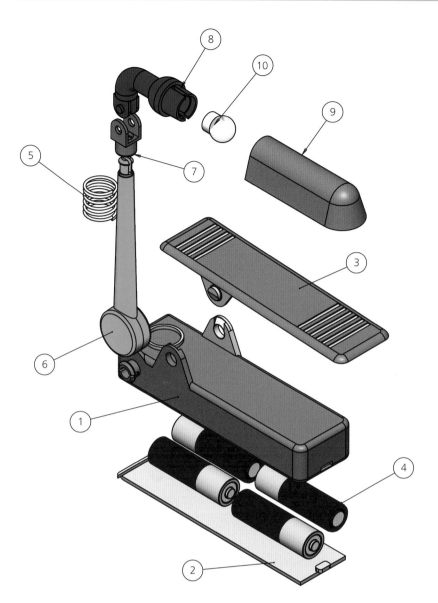

ITEM NO.	PART NAME	QTY.
1	Base	1
2	Cover	1
3	Clamp	1
4	Battery	4
5	Spring	1
6	Arm	1
7	Swivel	1
8	Elbow	1
9	Shade	1
10	Bulb	1

Figure 2.37 Exploded isometric view of a reading light with numbered parts list

Further labels on the drawing identify the head and lens of the torch, while annotations expand on this and specify that the bulb is a high-brightness xenon bulb and the reflector is metallised plastic (plastic with a metal surface finish). While labels are used to identify components by number or using simple text, annotations provide a more detailed explanation of the component or feature. An annotation could explain how a component is to be made, its quality of finish or how it operates.

Labels, annotations, parts lists and details of materials can be added to both 3D and 2D engineering drawings.

ITEM NO.	QTY.	PART NAME	MATERIAL
1	1	Base	Polystyrene
	1	Holder	Polystyrene
	1	Round swivel cap	Polystyrene
	2	Battery AA	Various
	1	Battery cover	Polystyrene
	1	Switch	Polystyrene/brass contacts
	1	Clip	Stainless steel
	1	Pin	Stainless steel
2	1	Swivel	Polystyrene
3	1	Head	Aluminium
4	1	Miniature bulb	Various (glass, metals)
5	2	Locking pin	Aluminium
6	1	Swivel clip	Polystyrene
7	1	Reflector	Metalilsed polystyrene
8	1	Lens cover	Aluminium and acrylic

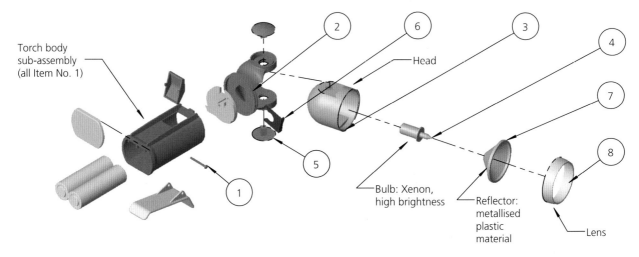

Figure 2.38 Swivel-head torch assembly with parts list, materials, labels and annotations

Activity

Produce a 3D isometric assembly drawing of a 13-amp plug, like the one shown in Figure 2.39. Practise producing the drawing by hand.

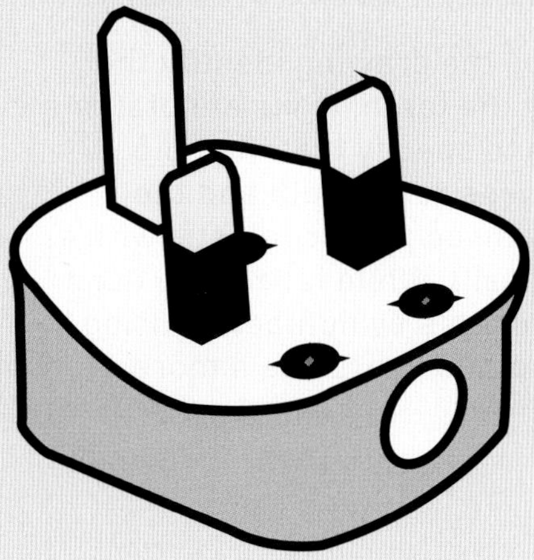

Figure 2.39 3D isometric assembly drawing of a 13-amp plug

Stretch activity

Draw an exploded view of a 13-amp plug. You might need to (safely) disassemble a plug to be able to do this. Make sure each part is labelled and include a tabulated parts list, showing part number, part name and quantity. You could also annotate with more detail the key components used to manufacture the plug.

2D engineering drawings

A 2D engineering drawing is a technical drawing which defines very accurately the requirements for engineered components and products. The drawing communicates clearly and accurately all the geometric features of the component or product, enabling it to be manufactured successfully.

While 3D drawings provide a pictorial view of the component or product, 2D engineering drawings are able to show detail. They are drawn to national and international standards to ensure they are correctly interpreted. In the UK, BS 8888 is the standard for engineering drawing. The standard is detailed and includes standard drawing conventions for many different types of engineering drawing techniques (such as scale and tolerance, dimensions, line types, abbreviations, mechanical features).

Synoptic link

We have already covered some of standard drawing conventions in Unit R038. Several of the most relevant are illustrated throughout this section – see if you can spot them.

Third angle orthographic drawings

An **orthographic drawing** represents a 3D object by using several 2D views of it. Typically, an orthographic drawing shows three different views (sometimes called **projections**), from the front, side and top of the object, as shown in Figure 2.40. This is known as **third angle orthographic drawing**. The side projection is usually the right-hand side. Occasionally, more views are shown if more clarity is required.

With orthographic drawing, all projections are always drawn on the same sheet, accurately aligned so that features match up and are drawn to the same scale.

The symbol (circles and cone) shown at the top of Figure 2.40 indicates that it is a third angle orthographic drawing. Third angle projection is commonly used in most countries, including the UK. However, in the USA, first angle projection is preferred, where the views are arranged differently, with the left-hand view being preferred to the right-hand view.

Figure 2.40 also includes accurate dimensions (in millimetres), enabling the component to be manufactured. Note the use of different line types – leader lines showing where dimensions are applied, dimension lines with arrow heads and **centre lines** showing the exact centre of the holes through the object.

Key terms

Orthographic drawing Drawing that represents a 3D object by using several 2D views (or projections) of it.

Projections 2D views of an object used to represent it in 3D.

Third angle orthographic drawing Orthographic drawing of an object which usually shows the front, right-hand side and top views.

Centre lines Lines drawn to indicate the exact centre of a part; always drawn using a series of shorter and longer dashes.

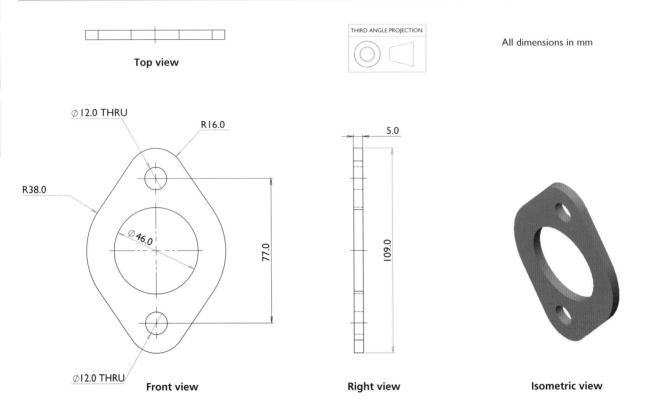

Figure 2.40 Orthographic drawing including an isometric view of the component

Dimensions, scale and tolerances

Dimensions are numerical values added to engineering drawings to communicate the sizes of key features. This enables components and products to be manufactured accurately.

Figure 2.41 shows a third angle orthographic drawing complete with dimensions. Engineering drawings usually show dimensions in millimetres and use leader lines and arrows to indicate to which part the dimensions correspond. This figure also includes two holes (with centre lines), and their dimensions are defined by their diameters.

It is important to include enough dimensions to describe all features, but not so many that the drawing becomes confusing.

Each projection of the block in each of the three views (front, side and top) is drawn to scale to give an accurate representation. While the engineering drawing will always include details of the scale used, it is important not to scale

(or measure) dimensions off the drawing that are not directly dimensioned with lines, arrows and values. This is to avoid inaccuracies in manufacture.

The drawing of the block in Figure 2.41 includes another drawing feature called a tolerance (which was mentioned in Unit R038). It is difficult, time-consuming and expensive to manufacture components with the exact dimensions shown on the drawing. However, it is also not always strictly necessary to be exact.

Tolerances show the amount of variation allowed in a given dimension of the manufactured component – that is, the maximum and minimum acceptable dimensions. In the case of the block in Figure 2.41, the length of the block can be 80 mm + 0.15 mm, or 80 mm – 0.15 mm. This means that the length can be between 79.85 mm and 80.15 mm and still be considered accurately manufactured. Tolerances can also be globally applied to every dimension on the drawing, and therefore are not always shown alongside each dimension.

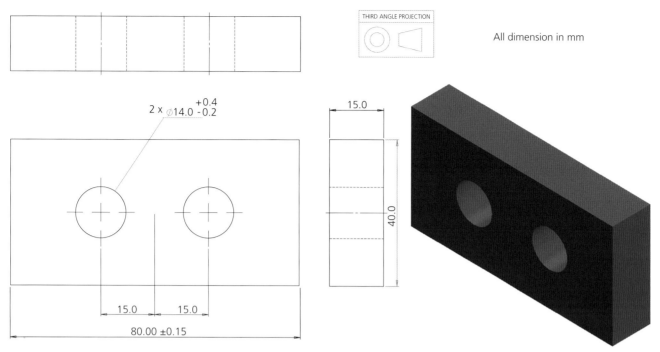

THIRD ANGLE PROJECTION

All dimension in mm

$2 \times \emptyset 14.0 \begin{smallmatrix} +0.4 \\ -0.2 \end{smallmatrix}$

15.0

40.0

15.0   15.0

80.00 ±0.15

Figure 2.41 Drawing of a block with dimensions and tolerances

Activity

Using a piece of A4 or A3 paper, draw a third angle orthographic drawing of the chair shown in Figure 2.42 (or a similar one).

Pick a suitable side of the chair to be the front and remember to include all three views – front, right and top.

Try to add some realistic dimensions to your drawing.

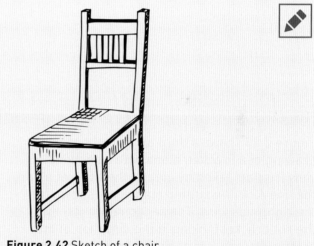

Figure 2.42 Sketch of a chair

Sectional views, drawing border and title block

In an engineering drawing, a **sectional view** (or sectional elevation) is used to show a 'cut-away' cross-section of an object so that the internal features can be detailed. It is done by removing or cutting away a section of the drawing (rather like cutting a cake). The cut is made along a line called the **cutting plane line**.

Key terms

Sectional view View used on an engineering drawing to show a 'cut-away' cross-section of an object so that the internal features can be detailed.

Cutting plane line Line showing where an imaginary cut is made through an object to expose a sectional view.

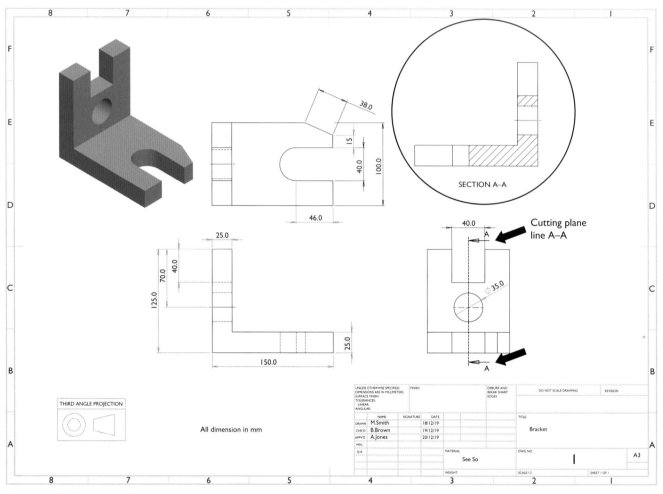

Figure 2.43 Orthographic engineering drawing of a bracket with a sectional view

Figure 2.43 shows a third angle orthographic drawing of a bracket, along with a sectional view (circled) taken by cutting the bracket along the line A–A (the cutting plane line) shown on the right-side view. The sectional view shows the cut bracket looking in the direction of the arrows on A–A. It is shown shaded (with **hatched lines**) where an imaginary knife cuts through the object, and plain white where the knife does not touch the object at all. See if you can identify where the hole through the bracket is shown on the sectional view.

Sectional views can be taken through any plane of the object, so long as the cutting plane line is identified to the reader. It is even possible to take a sectional view of a 3D drawing, such as from an isometric drawing.

The drawing of the bracket is shown with a **drawing border** – a line around a drawing that delineates the design area and often provides grid references to easily identify areas of the drawing. For instance, the circled sectional view is located in the area E2 to E3.

Key terms

Hatched lines Diagonal parallel lines which show where an object has been cut.

Drawing border Line around a drawing that delineates the design area and often provides grid references to easily identify areas of the drawing.

Unless otherwise specified, dimensions in millimetres (mm)				Tolerance: Over 6 up to 30 ± 0.2 Over 30 up to 120 ± 0.3	Deburr sharp edges	DO NOT SCALE		Revision: 1
	Name	**Signed**	**Date**			**Title:** Bracket		
Drawn	M. Smith	MS	18/12/19					
Checked	B. Brown	BB	19/12/19					
Approved	A. Jones	AJ	20/12/19					
					Material: ABS	**Drawing no.** 1		A3
					Weight:	**Scale:** 1:2		**Sheet 1 of 1**

Figure 2.44 A title block on an engineering drawing

In the bottom right of the border is a **title block** (A1 to A4), shown enlarged in Figure 2.44. The title block includes many details about the drawing, including the title of what is being shown (the bracket), the revision number of the drawing, the drawing size and who drew and checked the drawing, along with the date. Information about the drawing scale is also included, although a warning 'Do not scale' is often included, meaning dimensions cannot be measured directly off the drawing itself. Further details such as tolerances, materials and finishes can also be included in the title block.

Research

Find out more how to create accurate engineering drawings, including how to use standard drawing conventions.

A good starting point is the MIT OpenCourseWare, which is a free, web-based, publicly accessible collection of teaching and learning materials. Search for 'Design Handbook: Engineering Drawing and Sketching' on the MIT website.

Key term

Title block Box, usually included in the bottom right corner of a drawing, that includes important information to enable the drawing to be interpreted, identified and archived.

Activity

Produce a third angle orthographic drawing of the electric toaster shown in Figure 2.45 (or of your toaster at home if you have one). The drawing should include a drawing border and completed title block. Try to include a section view.

Add dimensions in millimetres for all the key features of your engineering drawing, using realistic values. You might need to undertake some further research to find these out. Remember to include the scale used.

Figure 2.45 Sketch of an electric toaster

Test your knowledge

1   What are oblique and isometric drawings, and what angle to the horizontal is typically used for each?
2   What are assembly drawings and exploded views?
3   What information is contained in a bill of materials?
4   What is a third angle orthographic drawing and which views are usually included?
5   How are dimensions correctly added to drawings, and what is meant by scale?
6   What does a section view of a drawing show?
7   What is included in a title block?

Assignment practice

Marking criteria

Mark band 1: 1–4 marks	Mark band 2: 5–8 marks	Mark band 3: 9–12 marks
Produces a **basic** orthographic drawing.	Produces an **adequate** and accurate orthographic drawing.	Produces a **comprehensive** orthographic drawing.
Produces an assembly drawing that is **limited** in detail.	Produces an assembly drawing with **some** detail.	Produces a **fully** detailed assembly drawing.
Production of drawings is **dependent** upon assistance or help from other sources.	Drawings are produced with **some** assistance or help from other sources.	Drawings are produced **independently**.

Top tips

- For this unit, you will be given a set assignment brief containing a scenario and tasks. Read this carefully and make sure you address all the points in the marking criteria. The scenario will be based on a product.
- You only need to fully develop one of the design ideas that you produced through sketching in the assignment for Topic area 1 of this unit.
- To fully satisfy the marking criteria, you need to produce both orthographic and assembly drawings. You could also include section views.
- Try to produce the drawings as accurately as possible using standard drawing conventions. Include sufficient dimensions so that it is possible to attempt manufacture using your drawings.
- Don't forget to include a parts list for assembly drawings, and suitable labels and annotations; you could also add a drawing border and title block to your drawings.

Model assignment

Scenario

A new bike light, like the one shown in Figure 2.46, is to be designed with the following requirements:

- battery-powered
- easy access to batteries
- high brightness but low power
- lens to intensify and protect the light source
- incorporate an on/off switch
- securely attached to the bike, but easy to attach and remove
- strong and robust casing
- easy to assemble
- aerodynamic design styling.

Figure 2.46 A bike light

Your task

You have already produced several initial design ideas using sketches. You are now required to develop one of your design proposals for the bike light. You should present your design solutions using engineering drawings.

Example candidate response

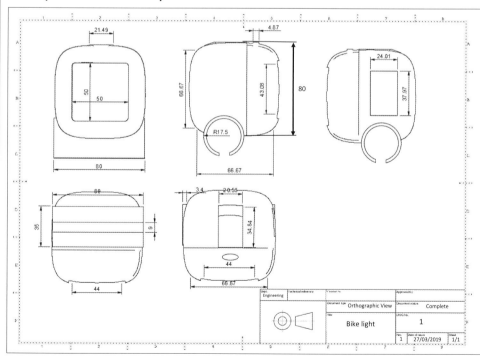

On the left I have given multiple orthographic working drawings, of the bike light, from multiple elevations. I have also included the basic dimensions of various different components so you can easily and clearly see how big each component is in proportion to the overall bike light.

These dimensions help to visualise the overall design of the bike light and how each component looks from a different angle, and how each component lays in relation to another.

The addition of basic measurements are also necessary for the bike light to be produced and manufactured accurately.

The candidate has selected a design for the bike light that has previously been produced through sketching and developed this into an orthographic drawing.

The candidate has drawn multiple projections, which are aligned, but this is not strictly a third angle orthographic drawing, as the front, right-side and top views are not in the correct positions.

The drawing is dimensioned and is in a drawing border with a completed title block.

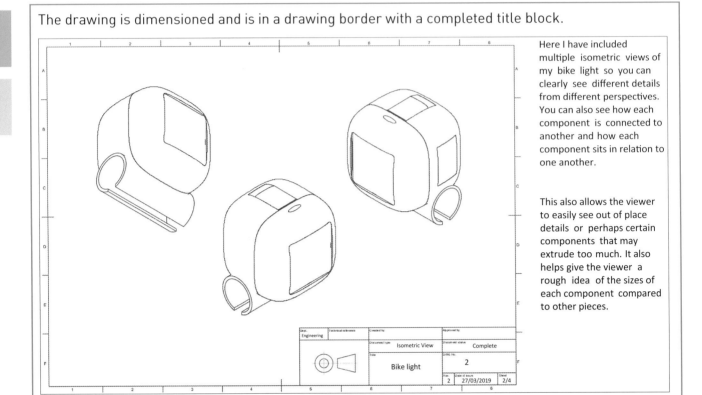

Here I have included multiple isometric views of my bike light so you can clearly see different details from different perspectives. You can also see how each component is connected to another and how each component sits in relation to one another.

This also allows the viewer to easily see out of place details or perhaps certain components that may extrude too much. It also helps give the viewer a rough idea of the sizes of each component compared to other pieces.

This drawing shows various 3D isometric views of the bike light and is useful for showing the finished product more clearly.

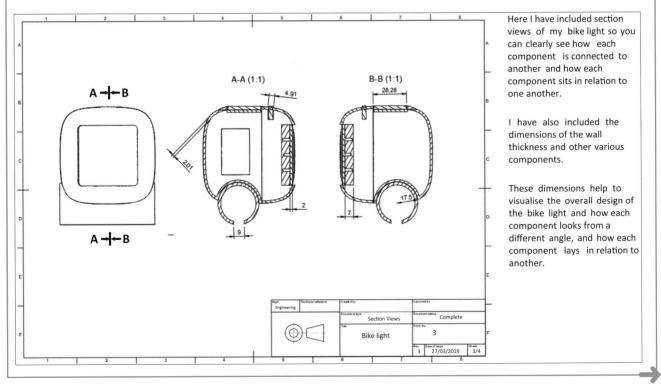

Here I have included section views of my bike light so you can clearly see how each component is connected to another and how each component sits in relation to one another.

I have also included the dimensions of the wall thickness and other various components.

These dimensions help to visualise the overall design of the bike light and how each component looks from a different angle, and how each component lays in relation to another.

The section view shows the bike light cut along the cutting plane line A–A and B–B. The line is in the same position for both, with the section view along A–A looking right, and that for B–B looking left. The section views clearly show the inside of the bike light, including the lens and bracket to attach it to the bike.

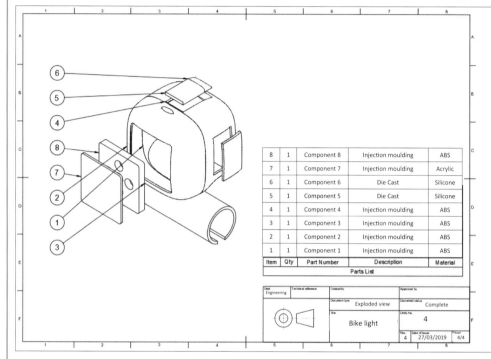

On the left, I have included an exploded working drawing of my bike light so you can quite clearly distinguish each component and how they are all assembled and connected to one another. In the bottom right hand corner, I have also included a numbered list of each component and the material and process that makes each component.

This exploded view helps a person to clearly visualise the overall finished product of the bike light. It is also incredibly helpful in the fact that you can clearly see the shape of the separate components that need to be produced, and how these components are connected to other pieces.

These pieces of information are a necessity when designing a product.

The exploded isometric view of the bike light shows clearly how it is assembled. Each part is numbered, and a bill of materials shows the quantity, materials and manufacturing process for each component.

Overall, the candidate has demonstrated the use of 3D drawing techniques (isometric and exploded view) and 2D drawing techniques (orthographic and section view). They have also included further engineering drawing features, such as dimensions, labels, drawing border, title block and bill of materials. While there are no drawing annotations, a separate commentary is provided for each drawing.

Topic area 3 | Use of computer-aided design (CAD)

Communicating design ideas and proposals, whether to the client or to potential consumers, is a key consideration for any designer. Methods of presenting designs for comment and feedback include building simple physical models, producing posters and display boards, and using presentation software like Microsoft PowerPoint.

However, specialist CAD software is now the most popular way to produce and present detailed 2D and 3D design ideas. While images and drawings can be produced quickly and efficiently to visualise and develop design ideas, using the software requires some practice and skill.

Getting started

Undertake some quick research to find out the benefits and disadvantages of using CAD software for producing engineering drawings compared to producing drawings by hand.

3.1 Produce a 3D CAD model of a design proposal to include compound 3D shapes

There are many different software packages that allow designers to produce professional engineering drawings (sometimes called draughting). Some packages are free and some packages require a licence fee. Nearly all enable both 2D and 3D engineering drawings to be produced quickly and accurately once the user has mastered how to use the software. Learning how to use any software takes time and patience, but the rewards are well worth it.

With nearly all engineering CAD software, images are first created in 3D, which allows rapid visualising of parts and assemblies. Later, these 3D images are converted into 2D engineering drawings, from which actual components and products can be manufactured.

While this book does not provide a whole course on CAD, it highlights a few of the most common techniques used in CAD software packages.

Interface

The **interface** is the means by which a user interacts with software. A typical CAD interface is shown in Figure 2.47. Like other computer software, a computer mouse and keyboard are used. CAD software often requires a powerful computer processor with sufficient memory in order to run efficiently. A large computer monitor also makes things easier, especially when working with large and complex drawings.

In the CAD software in Figure 2.47, commands are selected using on-screen menus or keyboard shortcut keys, with the drawing being produced in the large white area. This particular software provides a step-by-step history down the left-hand side of the screen as the part or assembly is created, for easy reference and modification.

Key term

Interface Means by which a user interacts with software.

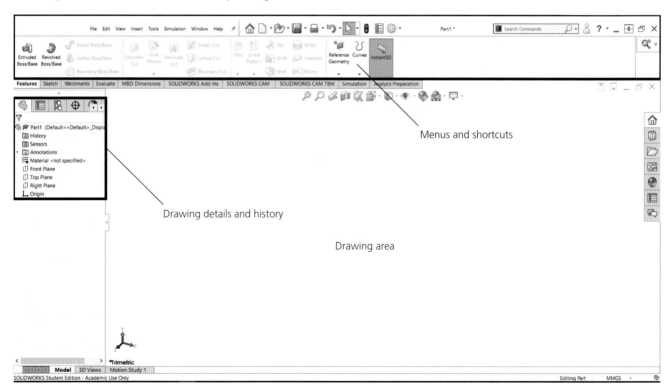

Figure 2.47 Typical CAD software interface

Sketch-based drawing

Basic sketch-based drawing tools, including lines, rectangles, circles and arcs, are used to create the geometry of parts. These are selected using one of the CAD menus, as shown in Figure 2.48. Further drawing tools include **polygons** and **splines**, which allow more complex shapes to be created, as shown in Figure 2.49.

You can use combinations of these drawing tools to create much more complex shapes, called **compound shapes**. As we will see, these can be used when using sketch-based tools to model parts to create complex 3D shapes.

Key terms

Polygon A plane (flat) figure with at least three straight sides and angles, and typically five or more.

Spline Smooth curve through a set of points.

Compound shape A shape made from a number of different shapes or elements put together.

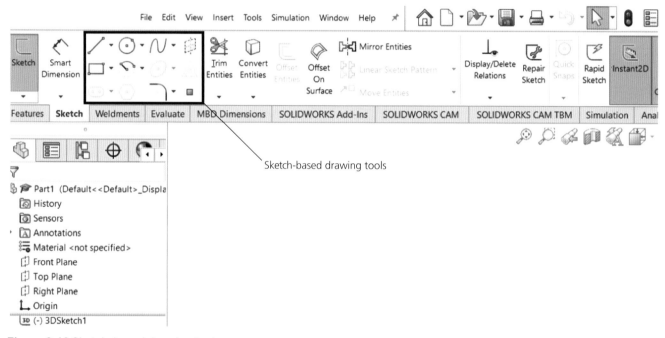

Sketch-based drawing tools

Figure 2.48 Sketch-based drawing tools

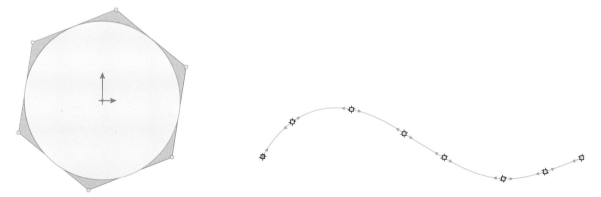

Figure 2.49 Polygon and spline being created

Solid modelling

CAD software uses a technique called **solid modelling** to model solid parts and assemblies in 3D space, which is controlled by accurately defining parts using their correct positions and dimensions using **reference geometry**. It means that parts need to be accurately aligned with **drawing planes** (sometimes called work planes), as shown in Figure 2.50.

There are three planes, called the front, top and right plane. These are at right angles to each other as shown, and use a coordinate system of x, y and z **axes** (shown by the coloured arrows in Figure 2.50). The point where the x, y and z axes meet is called the **origin**, with coordinates 0,0,0. It is important for the drawing to be aligned with the required plane, which can be achieved by rotating the planes and the drawing, and also to place the drawing on the origin correctly.

Key terms

Solid modelling Computer representation of a 3D solid object that can be used in design and simulation.

Reference geometry Defines the shape or form of a surface or a solid, and includes items such as planes, axes, coordinate systems and points.

Drawing (or work) planes Computer-based representations of 3D space.

Axes (singular axis) Reference lines for the measurement of coordinates – usually x, y and z in 3D space.

Origin Point where axes meet.

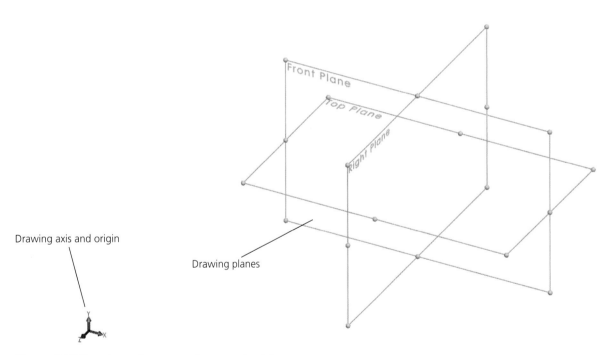

Drawing axis and origin

Drawing planes

Figure 2.50 Drawing using axes, planes and origin

Basic part modelling

In basic part modelling, the profile of the object is drawn and then the object is either rotated or stretched to create a solid object. The profile is rotated using the **revolve tool** and the profile is stretched using the **extrude tool**.

In Figure 2.51, the profile of a part has been created using simple lines. Note that the drawing is drawn flat onto one of the planes and the bottom left corner is placed on the origin. It is a compound shape as it is two rectangles joined together. It is important that sufficient dimensions are used to define (or describe) the profile, but not too many. A correctly dimensioned profile is termed **fully defined**, and the CAD software will warn the user if the profile is under- or over-defined.

Figure 2.52 Using the revolve tool to create a final part

A second part is being created in Figure 2.53, using the rectangle tool to first draw the basic profile of the object. Again, dimensions have been used to define fully the size of the profile, and the part has been drawn flat onto one of the drawing planes. Centre lines have been used to locate the point right in the middle of the rectangle for use in the next step. The origin is also located at this point.

The part is made into a solid block by using the extrude tool to pull the rectangle along an axis, as shown in Figure 2.54. To complete the part, a **hole** is cut into it using the cut tool, as shown in Figure 2.55. The hole is positioned at the point where the centre lines cross in Figure 2.53.

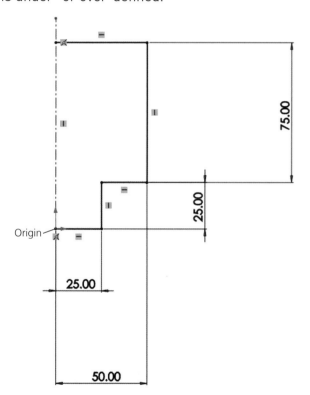

Figure 2.51 Using lines to construct the profile of a part

Figure 2.52 shows the profile rotated through 360 degrees using the revolve tool to create a solid part. The outline of the profile is clearly visible within the solid object.

Key terms

Revolve tool CAD tool used to create a 3D solid object by rotating a 2D profile around a centre line.

Extrude tool CAD tool used to create a 3D solid object by stretching a 2D profile sketch along an axis.

Fully defined Description of a correctly dimensioned profile.

Hole A hollow place or opening in a solid body or surface.

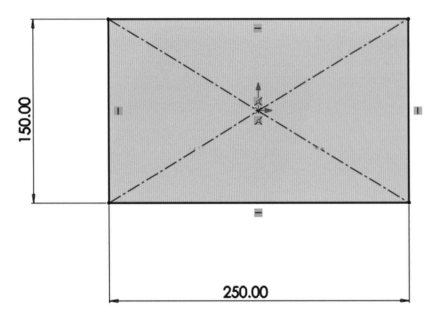

Figure 2.53 Using the rectangle tool to construct the profile of a part

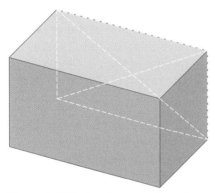

Figure 2.54 Using the extrude tool to create a block from a rectangle

Figure 2.55 Using the cut tool to cut a central hole into a block

Shell features

CAD software includes many more drawing tools to enable parts to be created easily. A particularly useful tool is the **shell tool**. This can be used to hollow out a part leaving open the faces selected and creating thin-walled features on the remaining faces, as shown in Figure 2.56. The thickness of the shell wall can be adjusted easily. It can also be used to create completely hollow components. This feature could be used to design casing, possibly to be manufactured by injection moulding.

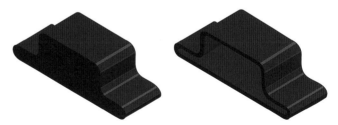

Figure 2.56 Solid part and shelled part

Key term
Shell tool CAD tool used to hollow out a solid part.

Creating assemblies

Once all the required parts have been created, the next task is to put them together into an assembly. Remember, all products are created by putting together separate parts and assemblies. No matter how complex the product or assembly, and the number of separate parts required, the process is always the same, with the parts being drawn first and then later assembled.

Figure 2.57 shows the two parts created previously ready to be assembled.

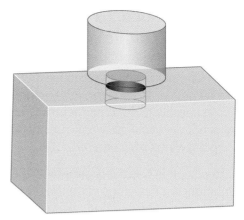

Figure 2.58 Two parts assembled precisely using the mate tool

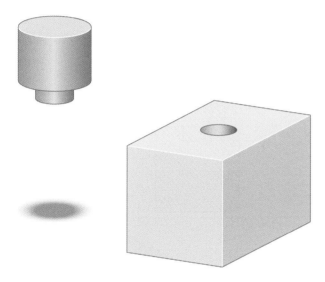

Figure 2.57 Two parts ready to be assembled

The CAD software allows the user to align parts accurately and to put them together using the **mate tool**. The tool often works by first aligning parts accurately, and then constraining (or fitting together) selected features of each component – often faces, edges, points or planes.

Figure 2.58 shows the two parts aligned and in the process of being mated. The parts are constrained and mated by selecting the appropriate features of each of the two parts which need to be fitted together.

Of course, in this example, it is important to ensure that when drawing the parts, the dimensions allow them to fit together.

Key term

Mate tool CAD tool used to align parts accurately and fit them together (sometimes called mate constraint).

Activity

A domestic appliance manufacturer wants to add rollers, like the one shown in Figure 2.59, to their products, to help when placing them under work surfaces.

Using CAD software, produce parts and an assembly of a suitable roller design. You can experiment with different sizes of roller and brackets to support the roller.

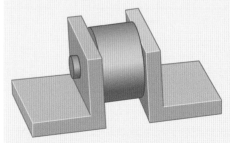

Figure 2.59 Roller assembly

Creating dimensioned 2D engineering drawings from 3D models

As we have seen previously, 3D drawings like those used to create parts and assemblies using CAD software do not convey enough information for the part or assembly to be manufactured. To manufacture products accurately and successfully, 2D engineering drawings are required.

With CAD software, it is relatively simple to create formal 2D engineering drawings once the 3D part or assembly has been created, as shown in Figure 2.60. If used correctly, the CAD software will use standard drawing conventions to produce these 2D engineering drawings. Further 2D drawings for each part could also be produced very simply. The drawing shows third angle orthographic views of the assembly and includes full dimensions to enable manufacture. An isometric view of the assembly is included for reference. A section view along A–A also shows the assembly sliced in half and reveals how the parts have been mated. It is drawn in a drawing template and includes a title block ready to complete.

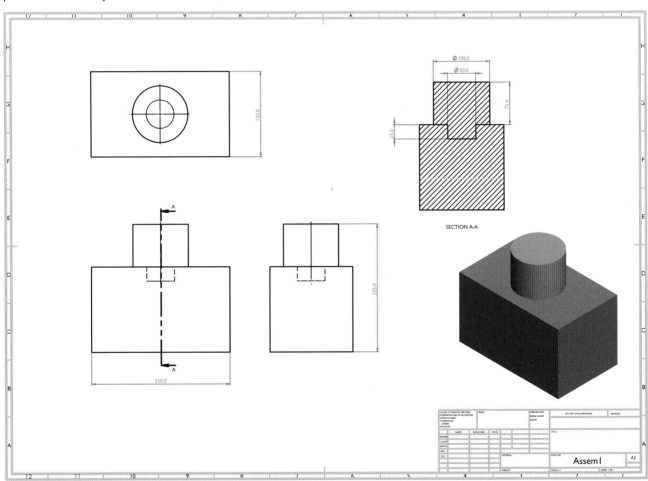

Figure 2.60 2D orthographic and 3D isometric drawing of an assembly

Activity

Produce 2D engineering drawings using the 3D models you developed for the roller assembly in the previous activity. You can create 2D drawings of the complete roller assembly and for each part separately.

Rendering and animation

CAD software includes many advanced features that enable parts and assemblies to be visualised. These include being able to add surface colour and rendering and being able to produce **animations**.

Figure 2.61 shows the assembly previously created rendered with a gold surface finish. The **rendering tools** within the software also add shade, including shadows, to the object by lighting it from different directions. Many different surface finishes and rendering techniques can be applied by the software, and a movie animation allows the object to be shown rotating. This is extremely useful to the designer when trying to convey design ideas to the client or consumer before actual production starts, and it allows design modifications to be made easily.

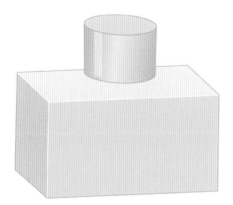

Figure 2.61 Rendered image of an assembly

Key terms

Animation Videos produced by CAD software for visualising products.

Rendering tools CAD tools that generate a realistic 3D image by adding features such as lighting, shade, reflections, tone and texture.

Simulation tools CAD tools that allow a model to be analysed virtually using engineering techniques and scientific calculations.

Stretch activity

Take the 3D model of the roller assembly you produced in the previous activity and use the surface colour, rendering and animation tools in your CAD software to create a rendered image to show to the manufacturer. Experiment with different surface colours and effects – for example, metals and plastics.

Advanced features

Some of the more advanced features of CAD software allow designers to prove that their designs can be manufactured successfully before producing a physical prototype. These are termed **simulation tools** and they enable the model to be virtually analysed using engineering techniques within the software. Examples include:

- motion studies showing how parts move in relation to each other
- simulation of stresses within the object
- simulation of the flow of air and liquids over the object.

Figure 2.62 shows a part created using CAD software undergoing a simulation of stresses within the object. Stresses are highest in the areas shaded orange and red, so the designer might wish to change the shape of the object to reduce these before manufacture, which can be done more simply at the design stage.

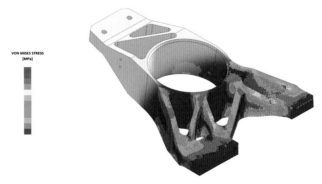

Figure 2.62 Simulation of stresses within an object

Research

Find out how to access the online tutorials for the CAD software that you are using. Here are a few useful starting points for some popular CAD software packages:

The AutoDesk website has free CAD software downloads for students and online tutorials: **www.autodesk.co.uk**

The Solidworks website includes tutorials on how to use their CAD software: **www.solidworks.com**

The Solid Edge website has free CAD software downloads for students and online tutorials: **https://solidedge.siemens.com/en/**

Case study

Save Our Planet Bottles want to produce a new range of eco-friendly refillable water bottles suitable for all ages, like the ones shown in Figure 2.63.

Using CAD, produce a final design proposal of a refillable water bottle to present to them. You might wish to start by exploring several different initial design ideas using sketches (see Topic area 1) before selecting one and producing a final design using CAD.

Use suitable engineering drawing techniques to communicate your design proposal, including producing 2D and 3D engineering drawings. You should also try using the animation and rendering tools in the CAD software.

Figure 2.63 Refillable water bottles

Test your knowledge

1 What is solid modelling, and why are the axes, origin and drawing planes important in CAD?
2 What are sketch-based drawing tools and how are they used to create compound shapes in CAD software?
3 What is a shell, and how is it created in CAD?
4 What are the functions of the revolve, extrude and cut tools when creating solid models of parts in CAD.
5 How can the mate tool be used to put together assemblies in CAD?
6 How can 2D engineering drawings be created from 3D models in CAD?
7 How are rendering and animation used within CAD software?

Assignment practice
Marking criteria

Mark band 1	Mark band 2	Mark band 3
Use of CAD to produce a **simple** model of the design proposal.	Use of CAD to produce an **adequate** model of the design proposal.	Use of CAD to produce a **complex** model of the design proposal.
A **simple** 3D virtual model consisting of a very limited number of components.	An **adequate** 3D virtual model consisting of some components.	A **detailed** 3D virtual model consisting of many components.
Production of a 3D virtual model is **dependent** upon assistance or help from other sources.	A 3D virtual model is produced with **some** assistance or help from other sources.	3D virtual models are produced **independently**.

Top tips

- For this unit, you will be given a set assignment brief containing a scenario and tasks. Read this carefully and make sure you address all the points in the marking criteria. The scenario will be based on a product.
- Remember to show step-by-step detail of how you have used CAD software to create your design proposals. You can do this by taking regular screenshots and annotating them to describe what you are doing. Don't forget to clearly identify your final designs!
- To satisfy the marking criteria fully, you need to demonstrate both 2D and 3D CAD modelling.
- Your design proposals could also include details of materials to be used, together with manufacturing processes and assembly methods.

Model assignment

Scenario

A new hot glue gun, like the one shown, is to be designed with the following requirements:

- be manufactured with a two-part moulding
- incorporate space for internal components
- incorporate a trigger to dispense the glue
- be easy to reload with a new glue stick
- be able to stand independently
- incorporate an area for branding
- be ergonomically comfortable to use
- be aesthetically pleasing.

Your task

Having produced several initial design ideas, you now need to develop one of your selected design proposals for the glue gun. You should use 2D and 3D CAD techniques to present final design proposals.

Example candidate response

The candidate has provided annotated screenshots to show how they used CAD software to design the separate parts for their glue gun.

Note that the shell feature has been used to hollow out the housing to allow components to fit inside the glue gun.

Alongside the 3D models, the candidate has produced 2D engineering drawings, including third angle orthographic views of each part.

The commentary justifies how the design proposals satisfy the design requirements – two-part design, space for internal components, trigger to dispense glue, and so on.

Materials, manufacturing processes and assembly methods for each part have also been considered.

Animated GIF

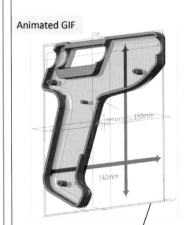

My first step to creating the glue gun on Creo was to first make the left housing of the entire model. I would have to create holes in the ABS housing for the details and components of the glue gun. I knew there would be an opening in the housing between the front and back housing to allow the glue stick to be inserted and loaded, so there is a semi circle gap in the back of the housing. This means that when put against the other side of the housing, there will be a fit circle for the glue stick, which later on will be presented in Creo too.

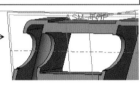

This animated GIF shows my final design of the hot melt glue gun, created in Creo. This GIF portrays how I built up the glue gun with all the components I designed from CAD and how I built it up during assembly.

I started with the left housing of the glue gun and slowly built it up by adding in the components one by one, the order of this being: nozzle, push lever, glue stick, cable, trigger, right housing and, finally, the metal stand of the glue gun so it can stand up by itself independently.

The other side of the glue gun has a lot of detail to it, that I showed cut out of the ABS. Some of this detail is for the look of the glue gun, but some of it is for reasons such as the logo space where the logo will be written. The other details are for the stand and the detail in the housing for sight of the glue gun and how it changes. These will be cut out of the ABS with a laser cutter because it's quick and easy to do so.

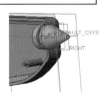

Here I have made sure that the left side of the glue gun housing has raised locator pins but the right side has indented locator pins. This means that the two sides can fit together perfectly without anything getting in the way, but it also means the two housings will stay together, along with these and its snap fits or screws. Snap fits would be ideal for DFMA because less material is being used, but screws will make sure it is easy for the user to open the glue gun up and fix it if a problem occurs.

The nozzle

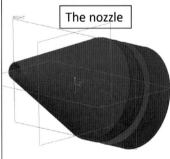

After the two housings were made in Creo, my next step was to follow the designs and create a nozzle for the glue gun. I would need to make sure it has the right dimensions to fit exactly into the glue gun housing without being loose or being too big to fit. I made the indented part of the nozzle 2.5 mm because that's how thick the ABS is on the housing of the glue gun, so it will definitely slot into the gap perfectly.

This is an engineering orthographic presentation drawing of the nozzle of the glue gun. It shows the design of the nozzle in 2D, with the dimensions of how big the nozzle will be. It shows how the nozzle will look from all views, and shows how it is designed with the gaps and spaces. The opening at the front of the nozzle will have a diameter of 25 mm, and the nozzle will be 30 mm tall.

The nozzle:
- will be made out of a cheap and easily manufactured metal such as brass.
- will have a total height of 30 mm, which isn't too big that it'll be heavy, plus it's saving as much material as possible, so I have thought back to R105 DFMA
- will get immensely hot when in use, so a rubber coating may be beneficial
- will slot into the front opening of the housing
- will be connected to an internal component called the heating element.

This part of the engineering drawing shows the dimensions for the nozzle of the gun. The gap where it will slot into the glue gun housing is 25 mm, as is the opening at the front of the nozzle where the glue comes out. The height of the nozzle itself is 30 mm.

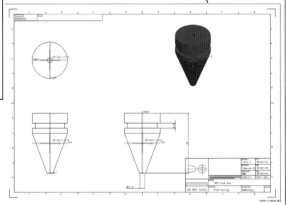

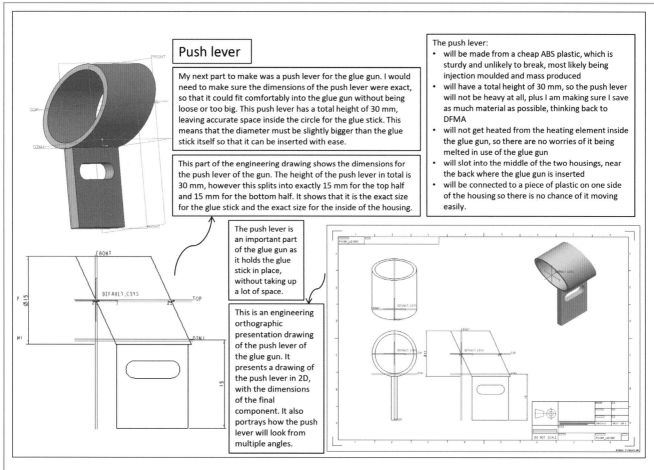

Push lever

My next part to make was a push lever for the glue gun. I would need to make sure the dimensions of the push lever were exact, so that it could fit comfortably into the glue gun without being loose or too big. This push lever has a total height of 30 mm, leaving accurate space inside the circle for the glue stick. This means that the diameter must be slightly bigger than the glue stick itself so that it can be inserted with ease.

This part of the engineering drawing shows the dimensions for the push lever of the gun. The height of the push lever in total is 30 mm, however this splits into exactly 15 mm for the top half and 15 mm for the bottom half. It shows that it is the exact size for the glue stick and the exact size for the inside of the housing.

The push lever:
- will be made from a cheap ABS plastic, which is sturdy and unlikely to break, most likely being injection moulded and mass produced
- will have a total height of 30 mm, so the push lever will not be heavy at all, plus I am making sure I save as much material as possible, thinking back to DFMA
- will not get heated from the heating element inside the glue gun, so there are no worries of it being melted in use of the glue gun
- will slot into the middle of the two housings, near the back where the glue gun is inserted
- will be connected to a piece of plastic on one side of the housing so there is no chance of it moving easily.

The push lever is an important part of the glue gun as it holds the glue stick in place, without taking up a lot of space.

This is an engineering orthographic presentation drawing of the push lever of the glue gun. It presents a drawing of the push lever in 2D, with the dimensions of the final component. It also portrays how the push lever will look from multiple angles.

The candidate has presented an assembled 3D model of the glue gun's final design, suitably rendered with colour and shading.

An animation of the glue gun is embedded into the presentation, which has been produced using presentation software.

The final presentation includes a summary of materials, manufacturing processes and assembly methods for the glue gun.

The candidate's work demonstrates a wide range of 2D and 3D CAD techniques, including: producing complex shapes; using revolves and extrudes; using the shell tool; using mates and constraints; creating virtual models; and using more advanced tools such as rendering and animation.

It also includes a summary of a class survey to identify the most popular colour for the design, along with suitably rendered examples to illustrate the different colour options.

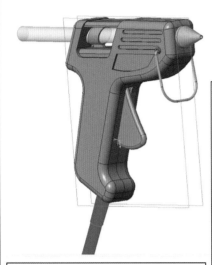

The visuals of my final design

The glue gun:
- will have these components involved: nozzle, push lever, glue stick, cable, trigger, stand and housing
- will have a housing 2.5 mm thick made from ABS plastic, making it as sustainable and durable as possible without being heavy
- will have components such as the nozzle and the stand made out of a metal like brass, which is easy to obtain and cheap
- will have an area for branding, where the logo could be placed
- will have a cable cover made from a rubber type material, which is a bad conductor of heat, making it safer
- will have accurate dimensions so everything fits together perfectly
- will have curved edges so it is safe for the user and they won't be able to cut themselves on sharp edges
- will be able to stand independently due to the metal stand
- will be aesthetically pleasing as the colours go well together and the glossy finish makes it look good.

This animated GIF shows how the glue gun will be assembled part by part. I have made this GIF by adding several photos of the glue gun I created and designed in Creo, each photo adding another component. This is the exact process of how I built up the glue gun during assembly. I started with the left housing of the glue gun and slowly built it up by adding in the components one by one, the order of this being: nozzle, push lever, glue stick, cable, trigger, right housing and, finally, the metal stand of the glue gun.

Animated GIF

This is my final design, made on Creo. I have created each component independently while taking into consideration the dimensions and how the glue gun will fit together. Then I put all the components together, using assembly in Creo. The animated GIF at the right hand side of the screen portrays how I put this together and how I got all the parts to fit together perfectly, while also showing how everything fits into the housing of the glue gun.

This slide shows how I have used Creo to change and alter the colours of my glue gun. I did this to see which colours would be the most aesthetically pleasing for the user. The original design, which was grey and light blue, went very well together and is very aesthetically pleasing for the customer. This is important because it means more customers will buy this glue gun over others due to the illusion of working better. I have, however, changed the light blue of the glue gun to colours such as yellow, pink and red. I think the yellow version of the glue gun is the most aesthetically pleasing out of the three because yellow and grey are quite complimentary.

Which glue gun is your favourite aesthetically?

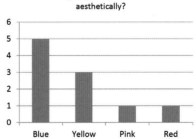

The hot melt glue gun will:
- be manufactured with a two-part moulding
- incorporate space for internal components
- incorporate a trigger to dispense the glue
- be easy to reload the glue stick
- be able to stand independently
- incorporate an area for branding
- be ergonomically comfortable to use
- be aesthetically pleasing.

Aesthetics can be important for a product, especially if it is a product that is electrical. If a product looks nice to a customer, they would most likely believe that this product works better than a worse looking one, even if this is not the case. It is also important because it allows the customer to believe they have spent their money well and have found something worth their money. This is why the looks of the glue gun would be important for the customers, also the specification says that the glue gun should be 'aesthetically pleasing'.

This graph shows which glue gun ten people preferred the aesthetics of, proving that the light blue glue gun is the most popular with five people, the yellow glue gun has three people who prefer it, and the pink and red glue guns both only have one person who preferred them. This means that the blue glue gun design is the one I will be taking forward to make a prototype of.

Synoptic links

Unit R039 allows you to apply the key knowledge, skills and understanding that you learned in Unit R038, particularly with reference to:

- the iterative design process and the generation of design ideas by sketching
- the communication of design outcomes using 2D and 3D engineering drawings and standard drawing conventions
- the use of CAD software, including its advantages and limitations compared with manual drawing techniques.

It also relates to manufacturing considerations that affect design.

The learning in Unit R039 enables you to read, understand and produce 2D and 3D engineering drawings using CAD, in preparation for creating a virtual CAD model and making a prototype in Unit R040.

Design, evaluation and modelling

About this unit

This unit provides an introduction on how to carry out product analysis and research. You will use product evaluation tools such as ACCESS FM and ranking matrices to compare the advantages and disadvantages of different products. You will also undertake a practical activity to disassemble and analyse a product. You will then create a virtual CAD simulation and use physical modelling to create a prototype, comparing this against a product design specification and identifying potential design improvements.

Topic areas

In this unit, you will learn about:

1 Product evaluation
2 Modelling design ideas

How will I be assessed?

This unit is assessed through an assignment that will take place at your centre. The assignment contains a scenario and a set of tasks for you to complete. Your work will be assessed against a set of marking criteria. The time to complete the assignment is included with the assignment brief.

Your teacher will provide you with clear guidance about the tasks required to complete the assignment, and the criteria which it is expected to meet.

Topic area 1 Product evaluation

Getting started

Working in small groups, choose an engineered product and discuss which of its features and functions are important for the consumer and why.

1.1 Product analysis

When designing new products, it is useful to research the design features of existing products and the requirements of consumers. Any findings can influence future designs. Before we can analyse any product, we first need to understand the different ways to do this using the different research methods available.

Key terms

Disassembly Taking something apart (for example, a product or piece of machinery).

Sample size Number of people included in a sample; it needs to be large and diverse enough to ensure the data represents the population.

Table 3.1 Primary and secondary research

Research type	Research methods/sources	Advantages	Disadvantages
Primary (field) research – gathering information that has not been collected before	<ul><li>Physical product analysis (product **disassembly** – taking things apart)</li><li>Interviews</li><li>Questionnaires</li><li>Surveys with consumers</li><li>Focus groups</li><li>Consumer trials</li></ul>	The information gathered is:<ul><li>relevant and up to date</li><li>specific to the researcher</li><li>only available to the researcher (unless made available to the public).</li></ul>	<ul><li>The process is costly and time-consuming to complete.</li><li>A suitable **sample size** is needed in order to be reliable.</li><li>Consumers are often unwilling to take part.</li></ul>
Secondary (desk) research – gathering information that has already been collected and published	<ul><li>Internet data</li><li>Books and trade literature (such as magazines and newspapers)</li><li>Manuals and brochures</li><li>Data sheets</li><li>Images and drawings</li><li>Standards and legislation</li><li>Government documents</li></ul>	<ul><li>The information gathered is often free.</li><li>The information is typically based on a large sample size and so is more accurate.</li><li>It is less time-consuming to collect data.</li></ul>	The information gathered:<ul><li>might not be up to date</li><li>is not specific to the researcher, and so might not include everything they need</li><li>is available to everyone, which may reduce any competitive advantage.</li></ul>

Research methods used for product analysis

As we learned in Unit R038, there are two main types of research:

- primary research – sometimes called field research
- secondary research – sometimes called desk or desk-based research.

For both primary and secondary research, the information that is collected can be either factual or based on opinion. Information and data based on numbers and measurements is called quantitative data. Information based on descriptions or observations which cannot be measured or counted is called qualitative data. While both types of data have their advantages and disadvantages, quantitative data is much easier to analyse and summarise.

> **Activity**
>
> Categorise each of the following activities as either primary (field) or secondary (desk) research methods:
>
> - analysis of a product through disassembly
> - manufacturer's web page
> - consumer questionnaire
> - maintenance manual
> - engineering technical drawing
> - focus group with consumers
> - British Standards for a product
> - consumer trial with a prototype.

Primary research methods

Physical product analysis

Physical product analysis involves researching and dismantling (taking apart) actual products. By taking a product apart, it is possible to find out:

- which components have been used
- which materials it is made from
- how it is assembled
- which manufacturing processes have been used
- how it can be maintained.

Physical product analysis can also include looking at the features of a product, and how it functions and operates. Analysis of this type is sometimes called **reverse engineering**.

> **Key term**
>
> **Reverse engineering** Taking apart and analysing a product's construction or composition in order to produce something similar.

The main benefits of physical product analysis are that real products can be analysed in detail, and that a range of products can be selected for analysis and the best features from each can be used to influence new designs.

However, there are also disadvantages. It can be time-consuming, expensive and difficult to perform. For example, it may be difficult to disassemble the product or to identify the materials and processes used.

Interviews

Interviews involve talking to people and asking them questions. The questions can either be:

- structured (closed) – for example, asking questions with a yes or no response, or asking people to rate something on a scale, such as from 1 to 10, or
- unstructured (open) – for example, 'What do you like best about this product?'

Structured questions are much easier to analyse and summarise than unstructured questions.

Interviews have several benefits:

- The information obtained is up to date and relevant.
- If needed, the person being interviewed can be asked to explain their answers.
- During the interview, further questions can be asked which were not originally planned.

However, undertaking interviews can take a lot of time and money compared to other methods, so they are not used in every situation.

Figure 3.1 Online questionnaires are a popular way to find out what customers think about products

Questionnaires

Questionnaires are a popular method for finding out what people think. They contain a series of questions which the participant answers and can be paper-based or now more commonly online (Figure 3.1).

As in interviews, questions can either be structured/closed (for example, yes/no questions, rate on a scale or place in order) or unstructured/open (where participants have space to answer freely). Again, structured questions are easier to analyse.

The benefits of questionnaires are:

- A large number of questionnaires can be distributed quickly.
- They are cheaper than other methods, such as interviews.

However, many people do not respond to questionnaires, and they need to be designed very carefully to get the questions right. It is also difficult to gain detailed responses to the questions or explanations for answers.

Surveys

A survey involves selecting a series of questions you want to ask and targeting a group who you want to answer the questions. Surveys usually involve using several different research methods. Interviews and questionnaires can all form part of the tools and techniques used to undertake a survey.

Surveys can be carried out face-to-face, by sending questionnaires in the post, by telephone and increasingly online. Often a mixture of some or all these methods is used. It is important to design the survey carefully to ask all the questions required, and to select how many people to ask (the sample size) to get reliable feedback.

Benefits of surveys include:

- The target group or groups (certain kinds of people) can be carefully selected.
- A range of different techniques can be used for the survey (for example, interviews, questionnaires).

However, there are also disadvantages:

- To get accurate results, the sample size needs to be sufficiently large and diverse.
- Large-scale surveys are time-consuming and expensive to undertake.

Focus groups

A focus group comprises a carefully selected group of individuals who are potential customers for a new product or who are users of similar existing products. This group is usually made up of a cross-section of the public, to ensure a wide range of views and opinions. Group members are invited to discuss their views and opinions on their wants, needs and preferences.

Figure 3.2 A focus group meeting to discuss new product development

There are several benefits of using focus groups:

- Group members can be carefully selected to be representative.
- Key questions and issues can be targeted.
- The information obtained is high quality, up to date and relevant.
- If needed, questions can be explained and discussed in detail.

- Other important issues may be highlighted by the group members, which the researcher may not have originally thought of.

However, there are also disadvantages. Undertaking focus groups can be:

- expensive
- time-consuming.

Consumer trials

With consumer trials, consumers are given sample products to try and asked for their feedback. The sample provided might be a pre-production prototype or even other competitors' products.

Benefits of consumer trials include:

- Consumers can provide genuine and honest feedback about a product before it goes into main production.
- The information obtained can be used to make modifications to the product before it is released onto the market.

However, consumer trials are expensive to operate, and analysing the views and opinions of people (qualitative data) is harder than analysing facts and figures (quantitative data).

Activity

A designer wants to use primary research to inform the design of a new electric toothbrush.

Working in small groups, discuss how each of the primary research methods could be used to do this.

Secondary research sources

Secondary research gathers and analyses information that has already been collected before. Therefore, it is vital to check how up to date this information is, and how this might affect its relevance to the research. The advantages and disadvantages of several secondary research sources are considered below.

Internet data

Internet data is information that can be found and accessed on the internet. It can be numerical facts and figures or text-based information.

Advantages of using internet data include the following:

- Many information sources can be found and searched very quickly.
- Information is global and can give insights into products and opinions in other countries.

However, there are also disadvantages:

- Information might not be up to date, so it is important to check when it was posted.
- The source of information might not be reliable, and information could be incorrect. It is crucial to check reliability.

Books and trade literature

Printed materials, such as books, magazines, trade journals and newspapers, offer a useful source of secondary information. Many of these can also be accessed electronically (for example, e-books and online newspapers).

Advantages of these sources of information include the following:

- They are often more accurate and reliable than internet sources.
- They are often more detailed than internet sources.

However, there are also disadvantages:

- It may be difficult or expensive to obtain some information, especially if it is only available in printed format (for example, books or journals).
- It is more difficult to search paper-based information.
- Information might not be up to date; it is important to check how current it is.

Manuals, brochures and data sheets

Manufacturers often publish user, installation and **service manuals**, as well as data sheets and sales brochures, for their products.

These sources of information are beneficial because they:

- can provide detailed technical information about a product
- give details of the product's features and functions.

However:

- Manufacturers might only publish some, not all, of the technical information, in order to protect their designs from being copied.
- Detailed technical data sheets and manuals are often restricted to service and maintenance personnel.

Key term

Service manual Document that provides step-by-step guidance on how to disassemble, maintain and reassemble a product safely.

Images and drawings

Images which provide detailed technical information about a product can be very useful when trying to analyse a competitor's products. Simple drawings of key parts of a product are often submitted with a patent and can be a good starting point when developing a new product (but remember that the design cannot be directly copied, as it is protected by the patent!).

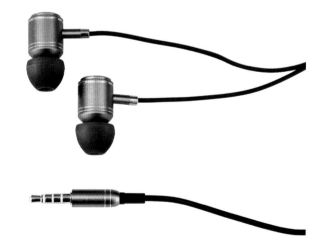

Figure 3.3 An image of ear buds can be used to research a competitor's product

Advantages of these sources of information include:

- Images can be used to investigate the features, finish and aesthetics of products.
- Detailed technical drawings provide all required detail of the product, such as key components, dimensions, materials, finishes and tolerances. They may also include manufacturing details.

Disadvantages include:

- Images need to show different views of the product to be useful.
- Like manuals, technical drawings are often restricted to certain personnel.

Standards and government documents

As seen in Unit R038, standards provide detailed requirements for the design of products, to ensure they are safe and reliable. They are a useful source of information and crucial to the design process.

Government publications can also provide useful information – for example, census data (a profile of the population), statistics about UK industry and guidance on sustainability targets.

These sources of information are beneficial for the following reasons:

- Definitive documents like standards provide key design considerations for products.
- Government publications are based on extensive expert research, accurate data and very large sample sizes.

However:

- Sometimes it can be expensive to access these sources of information (for example, it costs to download standards from the BSI website), although some are freely available.
- It takes skill to be able to read and understand documents correctly.

Activity

Working in small groups, discuss the advantages and disadvantages of using secondary research when producing a new design for an electric kettle.

Analysing and comparing products

Many primary and secondary research methods can be used to investigate and review other manufacturers' products, in order to inform the development of a new product. This needs to be done in an organised way, so that it is possible to compare the features, functions and detailed design across a range of products and select the best features from each.

There are many ways to analyse and compare products. In this section, we will show you how to do this practically using ACCESS FM, ranking matrices and a more detailed approach called quality function deployment (QFD). The examples show a comparison of different mobile phones, like the ones shown in Figure 3.4.

Figure 3.4 A selection of mobile phones

ACCESS FM

ACCESS FM was introduced in Unit R038. It is used to analyse products systematically and to compare the relative strengths and weaknesses across a range of similar products.

Table 3.2 shows ACCESS FM being used to analyse a mobile phone. Typical questions used to complete the analysis are also included against each heading. Note that a table is used to present and summarise the data so that it can be easily interpreted and analysed. You might think of more detail to add to the analysis.

Table 3.2 ACCESS FM analysis of a mobile phone

ACCESS FM	Typical questions	Mobile phone analysis
Aesthetics	• What does it look like? • Describe the texture, shape, colour, and so on. • Is it a modern or a traditional style? • Does it have a quality finish? • What has influenced the design?	• Sleek and slim design, gloss black with rounded corners. • Large glass screen and minimal buttons with modern design styling. • High-quality finish, with design influenced by other market-leading mobile smart phones.
Cost	• What is the estimated production cost? • What is the expected selling price? • What is the relationship between the two? • Is it affordable? • Is it good value for money? • Does the price reflect social and moral considerations (such as Fairtrade)?	• Production cost estimated at £100, with selling price of £540; selling price approximately five times the production cost so large profit made. • Quite expensive to buy, but good value for money given quality and features included. • Manufacturer claims to be an ethical trader, with good proportion of profit being passed on to overseas suppliers.
Customer	• Who would buy this product? • Who is the target market? • How does it appeal to them? • How would they use it? • What value does it add to their lifestyle?	• Would possibly be purchased by younger adults due to high selling price and complexity of features. • Style and features would be attractive, particularly for taking high-quality photos and accessing social media.
Environment	• What impact does the product have on the environment (consider the 6Rs of sustainable design from Unit R038 – recycle, reuse, repair, refuse, reduce and rethink)? • What are the impacts of manufacturing, use, distribution, packaging and disposal when the product is no longer needed?	• Product contains batteries and electronic components that would require safe disposal. • Possible to recycle and reuse some components. • Many components manufactured and assembled in other countries, which has an environmental impact on transportation of these.
Safety	• What safety issues and features have been considered? • Does the product meet safety standards?	• Includes password to restrict access to phone, protecting personal data. • Meets safety standards related to radio signals (health) and has the CE mark. • WEEE symbol included for safe disposal.
Size	• What is the physical size of the product? • Are there any other things that can be measured, like weight, area, volume, density and supply voltage? • Is the product comfortable to use (anthropometrics and ergonomics)? • Are the proportions appropriate for use? • If you increased or decreased the size of any features, would it work or look better? • How have human factors (such as ergonomics) been considered?	• 70 mm (wide) x 145 mm (long) x 9 mm (deep). • Weight 227 g; screen area 9100 mm^2; USB charger 5 V. • Fits well into an adult hand but might be too large for a younger person's hands. • Good screen area is useful to access software and view photos and apps. • Larger screen area might make phone easier to use.

ACCESS FM	Typical questions	Mobile phone analysis
Function	• How well does the product work? • Why does it work this way? • How easy is it to use? • How could it be improved?	• Works very well; quick response to touchscreen, which is easy to use. • Battery life could be improved, as only lasts one day. Better battery life would improve usability. • Internal camera OK but could be better quality. • Internal speakers are not that loud, so could be improved.
Materials	• What materials is the product made from? • Would different materials work better? • What impact does the choice of materials have on the environment? • How has the scale of production of the product affected the choice of materials and manufacturing processes?	• Case: injection moulded plastic; screen: glass. • Glass screen could be made scratch proof using a different type of glass. • Case and glass screen are easy to recycle. • Phone is mass produced, so injection moulding is a good manufacturing process to use.

The ACCESS FM analysis can be performed for many different types of product using the same basic headings, but not all questions need to be used. Both primary and secondary research can be used to find out the information required to complete the ACCESS FM analysis.

When looking at a range of similar products, like mobile phones, a separate ACCESS FM analysis would be done for each one and the results would then be used to compare the strengths and weaknesses across them using tools such as a ranking matrix.

Ranking matrices

While ACCESS FM and other methods are useful to perform an analysis of individual products, it is more useful to both the designer and the customer to be able to compare features and the strengths and weaknesses across a range of different products to inform their choices. Using ranking matrices is a useful way to compare products objectively. An example of a ranking matrix comparing three different mobile phones previously analysed using ACCESS FM is shown in Table 3.3. A scale of 1 (worst) to 10 (best) has been used to rank each of the phones against the ACCESS FM headings.

Table 3.3 Ranking matrix for three mobile phones

	Mobile phone A	Mobile phone B	Mobile phone C
Aesthetics	8	4	9
Cost	5	7	8
Customer	8	6	5
Environment	6	5	6
Safety	7	7	7
Size	7	3	9
Function	5	9	8
Materials	7	6	8

The ranking matrix can now be used to compare products by referring to the individual ACCESS FM score for each phone to be able to identify the strengths and weaknesses. For example, mobile phone B is clearly less aesthetically pleasing than mobile phone A and C and so this is a weakness for that phone. It does, however, have the best rating for function, which is a strength.

Another way to summarise the data more clearly in the ranking matrix is to use a bar chart, as shown in Figure 3.5. This makes comparison a lot simpler.

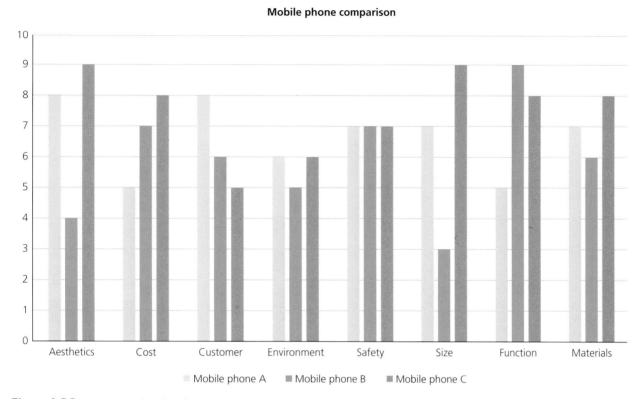

Figure 3.5 Data summarised and presented using a bar chart

Of course, the numbers used to rank products (from 1 to 10) in the ranking matrix need to be determined and this might be open to interpretation by an individual. While one person might score a 6, another might score a 3 or 7 under the same heading. Often, ranking matrices are completed through group discussion and agreement.

Quality function deployment

Quality function deployment (QFD) is a process used to make sure a product meets the needs of the customer. It tries to understand what their requirements and priorities are and to relate these to the technical requirements of the product. It can be used as part of the design process and planning for manufacturing. A QFD matrix for a mobile phone (and comparison with other mobile phones) is shown in Figure 3.6. The QFD matrix looks a lot like a house, and so it is often called the House of Quality.

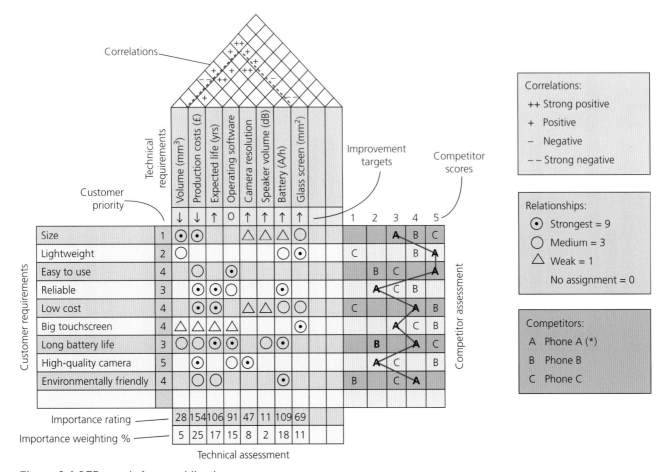

Figure 3.6 QFD matrix for a mobile phone

While the QFD matrix looks complex, it is fairly simple if done step by step. Here are the steps involved to build the QFD matrix for the mobile phone.

Step 1

Add the customer requirements and priority. In Figure 3.6, these are size, lightweight, easy to use, reliability, low cost, big touchscreen, and so on. Each one has a priority from 1 (least important) to 5 (most important) assigned to it. These have been determined using primary and secondary customer research.

Step 2

List the technical requirements for the product. For the phone, these are volume, production cost, expected life, operating software, camera resolution, and so on. Try to use requirements that can be measured, such as volume in mm³

and production cost in £. However, not all of them need to be measured.

Step 3

Add improvement target arrows to the technical requirements. For example, it is desirable for the product to have a lower production cost (down arrow), but the resolution of the camera should be higher (up arrow).

Step 4

Complete the centre grid to assign the relationship of the customer requirements to technical requirements. You will see that three different relationship symbols (circles and a triangle) are used, worth 1 point, 3 points or 9 points. As an example, if a customer wants a low-cost phone then this will affect the production cost, so the '9' symbol is placed in the box where they meet as there is a strong relationship. The big touchscreen is only weakly related to the operating software so a '1' symbol is used. Where they are not related, it is left blank.

Step 5

The next step is to calculate the importance ratings and importance weights along the bottom. The customer priorities and relationship symbol values are used to calculate the final rating values.

This is a little trickier to do, so here is the first column (volume mm^3) with four relationship symbols done in full.

Table 3.4 Calculating importance rating for first column in Figure 3.6

Customer requirements	Customer priority	Relationship symbol	Score
Size	1	⊙ = 9	1 × 9 = 9
Lightweight	2	◯ = 3	2 × 3 = 6
Big touchscreen	4	△ = 1	4 × 1 = 4
Long battery life	3	◯ = 3	3 × 3 = 9
Total:			28

All the other columns are completed in the same way. To work out the percentage, divide this total by the total of all the importance ratings:

$28 \div (28 + 154 + 106 + 91 + 47 + 11 + 109 + 69) = 28 \div 615 = 5\%$

Step 6

Complete the roof of the house, which is called the **correlation** matrix. Here + and – symbols are used to indicate the strength of the link between each of the technical requirements. In our example, there is a strong positive link between the volume (mm^3) of the phone and the area of the glass screen (mm^2) so a '+ +' is entered where they meet (shown by the red dashed lines). The battery capacity has no impact on the screen size, so '– –' is entered where these two meet.

Several of the others have been entered into the matrix.

Step 7

Nearly done! The final step is to complete the competitor assessment matrix on the right. In our example, the main QFD matrix is for phone A, and this has been compared against competitors' phones B and C using a scale of 1 (worst) to 5 (best) against each of the customer requirements. A solid red line shows how phone A performs.

The finished QFD matrix provides a complete picture to help with the design and manufacture of the mobile phone. We can easily see that having a high-quality camera is most important for the customer, and that production cost and the battery capacity are important technical requirements for manufacture. The competitor assessment shows that for phone A, while it is the most lightweight and easy to use, the camera is the worst, which is something to focus on in the design.

Like the simple ranking matrix done earlier, the way the QFD matrix is produced is open to interpretation by the individual so they are often completed as a group activity.

Stretch activity

Use the internet to find out about the features and functions of five different types of wireless computer mouse.

Include in your research facts and figures (quantitative research) and consumer opinions, such as those found in consumer reviews (qualitative research). Use ACCESS FM to analyse the products and summarise your findings in a table.

Case study

The designers at the Ultra Loud Music Company want to come up with design ideas for new wireless headphones. You have been asked to help them by researching existing wireless headphone designs using both primary and secondary research methods.

Working in small groups, use at least one primary research method and one secondary research method to investigate the designs of five different wireless headphones already available. Use appropriate methods to summarise and compare the strengths and weaknesses of each design.

Research

Use the internet to find out more about how to use ACCESS FM and QFD for product design and development. Try to find some examples of different products that have been analysed using these methods. A good starting point is YouTube, where you will find some interesting product case studies.

Test your knowledge

1 What is meant by primary research and secondary research?
2 What are suitable research methods for carrying out primary and secondary research?
3 How can the ACCESS FM method be used to look at the strengths and weaknesses of existing products?
4 How is a ranking matrix used to compare the strengths and weaknesses of products?

Assignment practice

Marking criteria

Mark band 1: 1–3 marks	Mark band 2: 4–6 marks	Mark band 3: 7–9 marks
Produces a **basic** product analysis of the key features of products using ACCESS FM.	Produces an **adequate** product analysis of the key features of products using ACCESS FM.	Produces a **comprehensive** product analysis of the key features of products using ACCESS FM.
Provides a **basic** description of the strengths and weaknesses of existing products.	Provides an **adequate** description of the strengths and weaknesses of existing products.	Provides a **comprehensive** description of the strengths and weaknesses of existing products.
Basic use of an engineering matrix.	**Appropriate** use of an engineering matrix.	**Effective** use of an engineering matrix.

Top tips

For this unit, you will be given a set assignment brief containing a scenario and tasks. Read this carefully and make sure you address all the points in the marking criteria. The scenario will be based on a product.

● Select the most appropriate research methods to investigate existing (similar) products to the one in the assignment brief; you should ideally look at between three and five different products.
● Use ACCESS FM to analyse the key features of the products and to describe their strengths and weaknesses. A suitable ranking matrix should be used to make a comparison.

Model assignment

Scenario

You have been asked to research a range of different bike lights, like the one shown in Figure 3.7, and to investigate how one has been designed and manufactured through performing safe disassembly.

Your task

Use ACCESS FM along with suitable primary and secondary research methods to identify the strengths and weaknesses of existing bike lights. Use a ranking matrix or quality function deployment (QFD) matrix in your comparisons.

Figure 3.7 A bike light

Example candidate response

The candidate has produced an ACCESS FM analysis for three different bike lights. They have also included images of the bike lights.

Bike Light 1

Aesthetics
Simple design with one button to turn on and off. Only available in black. It is not very aerodynamic although it looks modern. Has mounting bracket to fix to handlebars. Light can be quickly removed from the bracket (so not left on the bike and cannot be stolen). It does seem to be made OK.

Cost
The cost of the bike light is £15.99. This is an average price for a bike light of this type.

Customer
This type of bike light would be bought by a bike user who uses their bike for going to work or school. It would be used when dusk on bike lanes or roads. It would keep them safe by warning other road users they are there.

Environment
The bike light uses replaceable batteries, so has an impact on the environment (disposing of old batteries). It is imported from Europe and so needs to be transported to the UK. It comes in a lot of plastic packaging which requires disposal.

Safety
The bike light is marked with the CE mark with shows it meets European safety standards. It also has the WEEE symbol which shows it can be disposed of safely.

Size
It is 80 mm long x 50 mm wide x 50 mm deep (excluding the mounting bracket). It takes two AA batteries. It weighs 94 g which is average for a bike light.

Function
The bike light only has one operating function – on and off. It is easy to use. It is not very bright, and would not light the path or road very well. It can be easily removed from the mounting bracket though.

Materials
It is made from ABS and is injection moulded. The mounting bracket has a rubber type material so it grips the handlebars and wont damage them.

Bike Light 2

Aesthetics
This is the most complex design with three buttons. Available in black or green. It is very stylish and modern. It has an adjustable lens that can be focused into a beam, and a wrist strap to carry it when taken off the bike. It is a little difficult to remove from the bike as it has a retaining clip on the mounting bracket. The bike light is very well made.

Cost
The cost of the bike light is £99.99. This is very expensive for a bike light.

Customer
This type of bike light would be bought by a serious biker, like someone who does mountain or track biking.

Environment
The bike light uses rechargeable batteries, which have limited impact on the environment (as they can be charged up). It is manufactured in the UK, and so does not require transporting from other countries.

Safety
The bike light is marked with the CE mark with shows it meets European safety standards. It also has the WEEE symbol which shows it can be disposed of safely. It also has the BSI Kitemark showing it meets UK standards.

Size
It is 160 mm long x 40 mm wide x 40 mm deep (excluding the mounting bracket). It has a rechargeable 9 V battery inside. It is quite heavy at 210 g.

Function
The bike light only has many operating function – the brightness can be adjusted and it can be made to flash. It is a little difficult to use. It is super bright, and would light the path or track very well. It is a little difficult to removed from the mounting bracket, but this means that it does not fall off on rough tracks.

Materials
It is made from strong die cast aluminium and plastics. The mounting bracket is solid, and clamps to the handlebars. It does not have any protection, and would easily mark the handlebars.

Bike Light 3

Aesthetics
This is the simplest design with only a slide switch. Available only in black. It is the most traditional design, and look like an old fashioned bike light. The bike light is not very well made, and the quality of finish is poor.

Cost
The cost of the bike light is £9.99. This is cheap for a bike light.

Customer
This type of bike light would be bought by a bike user who only uses their bike occasionally in the dark. It would keep them safe by warning other road users they are there.

Environment
The bike light uses replaceable batteries, so has an impact on the environment (disposing of old batteries). It is imported from China and so needs to be transported to the UK. It comes in a cardboard box, which can easily be recycled.

Safety
The bike light is marked with the CE mark with shows it meets European safety standards. It does not have more safety markings.

Size
It is 100 mm long x 70 mm wide x 50 mm deep (excluding the mounting bracket). It takes four AA batteries. It weighs 150 g which is above average for a bike light.

Function
The bike light only has one operating function – on and off. It is very easy to use as it only has a slide switch. It is very dim, and would only warn other road users that the biker is there.

Materials
It is made from very thin ABS and is injection moulded. The mounting bracket is made out of very thin metal, and would easily damage the handlebars if incorrectly fitted.

It is not clear how the information has been obtained – from secondary or primary research.

The ACESSS FM analyses are quite detailed and include some good points though, some of which are objective, like sizes, weights and costs.

Comparison of strengths and weaknesses

Aesthetics
Bike Light 3 has the best aesthetics although it is the most expensive. Bike Light 4 is the worst looking bike light as it looks very old fashioned and not very modern.

Cost
While Bike Light 2 is the most expensive, it does have the best features and is well made. Bike Light 3 is the cheapest and would be OK for occasional use, so this is a strength. Bike Light 1 is average cost for a bike light.

Customer
Both Bike Lights 1 and 3 are suited to the same type of customer who will use them for warning other road users they are there. Bike Light 2 is for a serious biker who wants to use the light to light up the track or path.

Environment
The most environmentally friendly bike light is Bike Light 2 as it uses rechargeable batteries. It is also made in the UK. Bike Light 3 uses more replaceable batteries than Bike Light 1, and both are made in other countries. The packaging on Bike Light 3 is the most environmentally friendly.

Safety
All the bike lights have safety marking, although Bike Light 3 only has a CE marking.

Size
The biggest and heaviest bike light is Bike Light 2, but it is for serious bike users. Bike Light 1 is a little smaller and lighter than Bike Light 3.

Function
Bike Light 2 has the most functions. Bike Light 1 and 3 have the same functions, although Bike Light 1 is much brighter than Bike Light 3 which is an advantage.

Materials
Bike Light 3 is made from better materials, while Bike Light 3 is manufactured using poor quality materials.

	Bike Light 1	Bike Light 2	Bike Light 3
Aesthetics	6	8	4
Cost	5	8	3
Customer	5	8	4
Environment	5	7	4
Safety	7	8	4
Size	7	9	5
Function	7	9	5
Materials	6	8	4

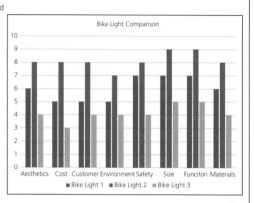

A ranking matrix has been used to score each of the ACCESS FM headings for each of the bike lights and a graph has been used to visualise the data. It is not clear if 1 is the best or the worst rating.

A reasoned comparison of the strengths and weaknesses under each of the headings has been produced.

Comparison of strengths and weaknesses

Summary

The best Bike Light is Bike Light 2 although it is the most expensive. It is for a serous bike user.

For a normal bike user, then the next best choice is Bike Light 1. It is good looking and well made and has a reasonable cost. It functions well to warn other road users, and can be easily removed from the bike.

For a very occasional bike user then Bike Light 3 would be OK. It is the cheapest but it is not very well made. It does function OK to warn other road users although it does look very good, and it is the least environmentally friendly.

	Bike Light 1	Bike Light 2	Bike Light 3
Aesthetics	6	8	4
Cost	5	8	3
Customer	5	8	4
Environment	5	7	4
Safety	7	8	4
Size	7	9	5
Function	7	9	5
Materials	6	8	4

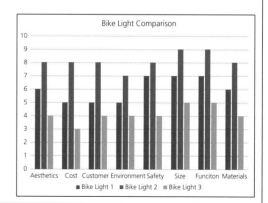

The candidate produces an overall summary, suggesting which is the best and which is the worst bike light, giving reasons why.

1.2 Carry out product disassembly

As we learned in in the previous section, a good way to find out how products have been designed and manufactured is to take them apart. By disassembling a product safely, it is possible to discover:

- which components it contains
- which materials have been used
- how it has been assembled
- which manufacturing processes have been used
- how it can be maintained in best working order.

The use of sources and procedures for disassembly

The easiest and safest way to disassemble (or service) a product is by following a set procedure. Sometimes product manufacturers provide a detailed service manual or guide, showing how to take their products apart to service them and how to replace parts. These include details of which tools are required, what safety precautions to take and a step-by-step disassembly and reassembly guide to be followed. They often include a series of annotated photographs.

Figure 3.8 is a workshop manual, showing step by step how to disassemble, service and reassemble parts of a Porsche 911. It includes detailed annotated photographs. This manual was published by Haynes, which independently produces manuals for a range of cars and other products.

Manufacturers often produce their own service manuals for larger or more complex items (for example, domestic appliances like washing machines or complex CNC machinery), although some are restricted to service personnel only (and not made generally available to users). This is to discourage users from servicing their own products, which might require special precautions, tools and skills.

Figure 3.8 An owner's workshop manual for a Porsche 911

For simpler, lower-cost products, there are often no service manuals available. One can be created in advance of taking the product apart, or sometimes more effectively alongside the disassembly process. Annotated photographs can be used to record each step and can then be used later to analyse the product in detail, and to enable it to be put back together again. There is a documented disassembly procedure in the practical disassembly task later in this section.

Disassembly procedures using appropriate tools and instruments safely

The first and most important thing to consider before undertaking any practical activity (such as dismantling a product) is how it is going to be done safely. Working safely is firstly about preventing injury to yourself or others, but also avoiding damage to equipment and property.

To understand how to work safely, we need to understand the terms **hazard** and **risk**:

- A hazard is something that has the potential to cause harm (whether to a person or property) – for example, the sharp blade on a knife.
- A risk is how likely a hazard is to cause harm – for example, the risk from the sharp blade of someone or something being cut.

Figure 3.9 A sharp blade on a craft knife is a hazard

Key terms

Hazard Something that presents a danger, either to physical health such as electricity or an open drawer, or to mental health.

Risk The chance or likelihood that someone could be harmed by a hazard and how serious the harm could be.

Table 3.5 Hazards and their associated risks

Hazard	Risk
Sharp craft knife	Being cut
Puddle of water	Slipping over
Electricity	Being electrocuted, causing a fire

Activity

Look at Figure 3.10 and identify the hazards and potential risks.

a)

b)

c)

d)

Figure 3.10 Hazards and risks

By knowing the hazards and risks associated with an activity, you can decide whether performing the activity is acceptable or whether there is some way you can reduce the likelihood of a hazard causing harm.

Of course, if the hazard is removed completely then there is no risk – for example, if we don't use a craft knife, we cannot get cut, or if the electricity is turned off, we cannot be electrocuted.

However, it is not always possible to remove the hazard; sometimes we will have to use the craft knife or work with the electricity turned on. In these cases, the best we can do is reduce the risk of harm, which can be done through taking extra care, by providing protection such as guarding or by wearing **personal protective equipment (PPE)**. These are called **control measures**.

Key terms

Personal protective equipment (PPE) Equipment designed to protect the user against risks to their health or safety.

Control measures Actions taken to reduce the risk (likelihood of a hazard causing harm), such as removing the hazard, taking extra care, guarding or wearing protective equipment.

Risk assessment Process of identifying, analysing and evaluating hazards and their associated risks, and seeing if they are acceptable or can be reduced.

Risk assessment

A widely accepted practice when undertaking any practical activity is to carry out a **risk assessment** before work commences. In a risk assessment, all the hazards presented by the activity need to be identified, together with the risks they present. The level of each risk is then considered and a decision is made about whether it is acceptable to continue with the activity. If not, then it might be possible to reduce the risk in some way so that the activity can go ahead more safely.

Risk assessment tables (like Table 3.6) are used to determine if the level of risk is acceptable. This is assessed in two ways:

- how likely it is that an accident will happen (likelihood), and
- how much damage or injury could occur if it does happen (severity).

The individual scores for each of these are then multiplied together to get the score shown in the box in the table. For instance, if an accident is likely (4) with moderate severity (3), the score will be 12. Ideally, we want to remain in the green area of the table and not have any risks in the red area.

Table 3.6 Risk assessment tables are used to determine if the level of risk is acceptable

		How much injury or damage could occur? (Severity)				
		1 Negligible	2 Minor	3 Moderate	4 Major	5 Catastrophic
How likely is it that an accident will happen? (Likelihood)	5 Almost certain	5	10	15	20	25
	4 Likely	4	8	12	16	20
	3 Possible	3	6	9	12	15
	2 Unlikely	2	4	6	8	10
	1 Rare	1	2	3	4	5

Here are two examples of simple risk assessments.

Sample risk assessment: bare electrical wires

Figure 3.11 Bare electrical wires

Table 3.7 Risk assessment: bare electrical wires

Hazard	Risk	Likelihood	Severity	Score
Bare electrical wires	Electrocution	4	5	20

As shown in Table 3.8, repeating the risk assessment with the wires insulated significantly reduces the risk, as the likelihood goes from 5 to 1.

Table 3.8 Risk assessment: insulated electrical wires

Hazard	Risk	Likelihood	Severity	Score	Control measures
Electrical wires (insulated)	Electrocution	1	5	5	Visually check wires are insulated.

Sample risk assessment: using a pillar drill

Figure 3.12 Using a pillar drill

Table 3.9 Risk assessment: using a pillar drill

Hazard	Risk	Likelihood	Severity	Score
Rotating chuck and drill	Long hair caught in the machine	4	4	16
Swarf (bits of metal) from drilling	Metal swarf in the eye	3	4	12

In this example, while getting hair caught in the machine or swarf in the eye might not prove catastrophic, they are still risks that need reducing. Implementing control measures significantly reduces the risks.

Table 3.10 Risk assessment: using a pillar drill with control measures

Hazard	Risk	Likelihood	Severity	Score	Control measures
Rotating chuck and drill	Long hair caught in the machine	2	4	8	Tie hair back Ensure guards in place
Swarf (bits of metal) from drilling	Metal swarf in the eye	1	4	4	Wear safety glasses

The likelihood of getting swarf in the eye is now significantly reduced; however, note that while the possibility of getting long hair in the rotating chuck is reduced, it is still recorded as 'unlikely', placing this hazard and risk in the amber area. Therefore, extra care must be taken for people with long hair.

Activity

Match each risk to its corresponding control measure.

Risk
Long hair caught in machinery
Loud noise from hammering operations
Cuts on hands from sharp edges
Dust from wood cuttings
Toxic fumes from soldering
Swarf from a milling machine

Control measure
Wear gloves
Tie hair back
Use a fume-extraction system
Wear ear defenders
Wear a dust mask
Wear safety glasses or goggles

Case study

Hot glue guns, like the one shown in Figure 3.13, are a popular way to join components together. They do, however, present some hazards with associated risks.

Working in small groups and using the template below, complete a risk assessment for using the glue gun safely. Identify hazards, risks and control measures.

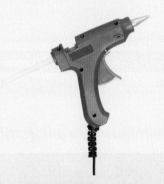

Figure 3.13 Hot glue gun

Hazard	Risk	Likelihood	Severity	Score	Control measures

While the purpose of undertaking risk assessments is to determine the risks of certain hazards and how these can be reduced, we may sometimes still have to carry out activities which fall within the amber or even red areas of the risk assessment table.

Producing risk assessments accurately is not simple; it is not easy to determine how likely the accident is to occur or its severity. While one person might score it a 1, another might score it a 2 or 3. For this reason, risk assessments are sometimes carried out by expert teams, and they are reviewed regularly.

The Health and Safety Executive (HSE) provides excellent guidance on its website (www.hse.gov.uk/risk) on how to produce risk assessments, including templates to complete.

Tools required

Safely dismantling a product requires tools. If a service manual or guide is available for the product, it will often list the tools required for its safe disassembly. Where no manual is available, it is up to the person dismantling the product to decide the best tools to use.

For simple products, relatively simple tools might be sufficient to dismantle it safely, such as:

- screwdrivers (for example, slotted and Phillips)
- Allen keys
- pliers
- side cutters
- spanners.

Figure 3.14 Slotted and Phillips (crosshead) screwdrivers

Figure 3.15 A range of different pliers and side cutters

Activity

Find out:

- what other types of screwdriver there are apart from slotted and Phillips
- what Allen keys and spanners are, and how they can be used to dismantle products safely.

Manufacturers sometimes use permanent fixings (such as rivets) or special temporary fixings which require special tools to remove. This is to prevent unauthorised disassembly by the user – especially if the product contains hazardous parts or materials or has no user-serviceable parts. Figure 3.16 shows tamper-proof screws, which are exceptionally difficult to remove.

Figure 3.16 Tamper-proof screws

Analysing an existing product through disassembly

We can learn a lot of things about the design and manufacture of a product by taking it apart.

Of course, when a product is disassembled and analysed, we can only use our best guess based on our knowledge to say how it was originally designed and manufactured. Without access to original detailed design information this can sometimes be tricky. Further research of similar products might confirm our initial findings from disassembly analysis.

Table 3.11 What we can learn from disassembling a product

Feature	What we can learn
Components	• Which components are used in the design and manufacture of the product • How many components are used • The dimensions of the components • The complexity of the design
Assembly methods	• How the product has been assembled • Whether permanent or temporary fixing methods have been used • Whether standard components have been used or fixings have been designed into the components themselves • How easy it is to take apart the product and reassemble it
Materials	• Which different materials have been used in the design and manufacture of the product • Why different materials have been used – for example, for their properties, robustness or aesthetics
Production methods and manufacturing processes	• Which production methods and manufacturing processes have been used to make the product (such as moulding, pressing and forming, material shaping, machining, finishing) • What scales of production could have been used to make the product (or parts of it) – one-off, batch or mass production
Maintenance	• How the product can be maintained (such as replacing or cleaning parts or checking things like fluid levels)

The process of taking something apart to reveal how it is designed and manufactured is often called reverse engineering.

Further information that we can discover through disassembly is how the product has been designed to make it easier and more efficient to manufacture and assemble. This technique is often used by designers working with manufacturers and is called **design for manufacturing and assembly (DFMA)**.

Key term

Design for manufacturing and assembly (DFMA) Process where a product is designed for efficient manufacture and assembly.

Stretch activity

Find out more about DFMA and how designers and manufacturers use this process to make the manufacture of their products more efficient. Refer to Unit R038, page 42, as a starting point.

Practical disassembly activity

In the final part of this section, we will perform the safe disassembly of a low-cost electric kettle. It includes:

● an initial risk assessment using a template
● step-by-step annotated photographic evidence of the disassembly
● evidence of the safe use of tools (in the photographs)
● a disassembly procedure produced alongside the disassembly
● detailed analysis of the kettle.

Risk assessment

The first step is to undertake an initial risk assessment of the activity. We might wish to add further items to the risk assessment if we discover additional hazards and risks during the disassembly.

Table 3.12 Risk assessment of the practical disassembly activity

Hazard	Risk	Likelihood	Severity	Score	Control measures
Electricity	Electrocution	1	5	5	Ensure the kettle is disconnected from the mains electricity
Screwdriver – pointed/sharp end	Puncture wound	3	3	9	Wear gloves
Pliers	Pinched skin	3	2	6	Wear gloves
Side cutters	Cuts to skin or clothing	2	3	6	Wear gloves
Flying sharp parts of kettle	Cuts to skin Damage to eyes	2	4	8	Wear gloves Wear safety glasses
Kettle components	Sharp edges – cuts Pinching in small gaps	3	3	9	Wear gloves
Possibly hazardous materials in kettle	Poisoning	1	4	4	Do not further disassemble sealed components – for example, the kettle element

Disassembly

Table 3.13 Disassembly of the electric kettle

Step, photograph and procedure/description

Step 1	Step 2	Step 3
Unpack kettle, kettle base and user instructions from box **Ensure kettle is not connected to mains electricity**	Gather PPE and tools required for activity: safety glasses, gloves, screwdrivers (slot head and Philips), pliers and side cutters	Turn kettle base over and remove screws (note use of gloves)
Step 4	Step 5	Step 6
Base with cover and base electrics removed Remove feet from base	Remove electrical cable from contacts (note safe use of pliers to remove spade connectors)	Kettle electrical cable and plug removed

Step, photograph and procedure/description		
Step 7  Remove screws from top of kettle to remove lid (note use of safety goggles)	Step 8 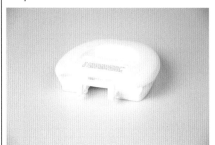 Remove kettle lid from body	Step 9  Remove screws from bottom of kettle to remove handle cover
Step 10  Remove screws holding kettle element, switch and contacts assembly to kettle body	Step 11  Kettle contacts and bi-metal switch assembly removed	Step 12  Bi-metal disc (which switches kettle off automatically when water is boiling)
Step 13  Heating element and waterproof seal to prevent leaks	Step 14  Heating element removed	Step 15  Remove indicator light from switch assembly by cutting wires using side cutters
Step 16 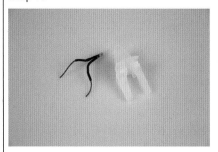 Switch lever and indicator light removed	Step 17  Completely dismantled kettle	

Analysis summary

Table 3.14 Analysis summary following disassembly of the electric kettle

Component	Materials	Assembly methods	Manufacturing processes/production methods	Maintenance considerations
Kettle body 	Polypropylene or polyethylene	Parts assembled to body with features built into moulding and screws	Injection moulded Batch or mass produced	Wipe clean with damp cloth
Kettle lid 	Polypropylene or polyethylene	Assembled to body using features built into moulding	Injection moulded Batch or mass produced	Wipe clean with damp cloth
Kettle base 	Polypropylene or polyethylene	Assembled using screws	Injection moulded Batch or mass produced	Wipe clean with damp cloth
Water level window 	Polypropylene or polyethylene	Glued or heat welded to kettle body	Injection moulded Batch or mass produced	Wipe clean with damp cloth
Feet 	Synthetic rubber	Push fit into holes in base moulding	Moulded Batch or mass produced	–

Component	Materials	Assembly methods	Manufacturing processes/production methods	Maintenance considerations
Kettle element	Stainless steel (casing only)	Fixed to base using screws	Various processes Batch or mass produced	Descale using kettle descaling chemical as required Replace if broken (if possible)
Element seal	Polymer (such as silicone, neoprene, EPDM)	Push fit into body moulding (has groove in seal profile)	Moulding or punching Batch or mass produced	Replace if damaged (if possible)
Thermostat	Bi-metallic disc: steel/copper or steel/brass	Disc is push fit into switch moulding assembly	Stamping and welding or brazing (bi-metal disc) Possibly mass produced	–
Electrical contacts	Brass	Securely fitted into switch moulding assembly	Extrusion or forging (could be machined) Mass produced	Clean if required
Electrical cable	Insulation: polyvinyl chloride (PVC) Wires: copper	Assembled to kettle contacts using push-fit electrical spade connectors	Insulation: extrusion Wires: 'drawing' (a type of extrusion) Mass produced	Check for damaged insulation
13-amp plug	Case: ABS or urea formaldehyde Pins: brass with nickel plating	Plug moulded onto cable – possibly standard component (plug and electrical cable assembly)	Case: injection moulded Pins: extrusion or forging (could be machined) Mass produced	Replace fuse if blown (check reason first)

Component	Materials	Assembly methods	Manufacturing processes/production methods	Maintenance considerations
Small parts 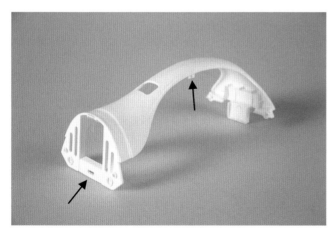	Self-tapping screws: bright zinc-plated steel	–	Various processes Screws: mass produced (standard component)	–
Packaging	Cardboard	–	Various processes Box: batch or mass produced	Recycle packaging

Further research was required to complete the analysis summary table, to find out what materials might be used for the kettle components, and their manufacturing processes.

Many injection-moulded components from the kettle have built-in features to allow them to be joined to other parts without the overuse of screws. In Figure 3.17, you can clearly see the slot in the bottom of the handle cover and other moulded features, which allow for quick and easy assembly. This is an example of DFMA and reduces the cost to manufacture and assemble the product. It also makes assembly less prone to errors.

Figure 3.17 The handle cover has been designed for easy assembly to the kettle body

Research

Find out more about how to work safely.

The Health and Safety Executive (HSE) website provides extensive guidance on risk assessments, including useful templates: **www.hse.gov.uk/risk**

Test your knowledge

1 How can a manufacturer's service manual help with safe disassembly, and what can be done if one is not available?
2 What are hazards and risks, and what method is used to see if risks are acceptable?
3 What are control measures?
4 How can a product be analysed through disassembly? Consider:
   ● components
   ● materials
   ● assembly methods
   ● production methods and manufacturing processes
   ● maintenance.

Assignment practice

Marking criteria

Mark band 1: 1–3 marks	Mark band 2: 4–6 marks	Mark band 3: 7–9 marks
Disassembly of a product is **dependen**t upon assistance or help from other sources.	Disassembly of a product is carried out with **some** assistance or help from other sources.	Disassembly of a product is carried out **independently**.
Limited understanding of potential hazards and safety considerations when using tools and equipment.	**Adequate** understanding of potential hazards and safety considerations when using tools and equipment.	**Clear** understanding of potential hazards and safety considerations when using tools and equipment.
Produces a **limited** analysis of the components, materials, production methods, assembly and manufacturing methods used in an engineered product.	Produces an **adequate** analysis of the components, materials, production methods, assembly and manufacturing methods used in an engineered product.	Produces a **comprehensive** analysis of the components, materials, production methods, assembly and manufacturing methods used in an engineered product.

Top tips

- For this unit, you will be given a set assignment brief containing a scenario and tasks. Read this carefully and make sure you address all the points in the marking criteria. The scenario will be based on a product.
- Before you start to disassemble and analyse a product for this part of the assignment, make sure that you will work safely – you should produce a risk assessment.
- If a service manual or other disassembly instructions are available, use them; if not, write your own brief instructions.

- Make sure you produce evidence of performing every step of the disassembly, which can be done with annotated photographs. This should also show you using tools safely and implementing any control measures.
- Make sure the analysis is complete, including components, materials, assembly methods, production methods and manufacturing processes. Don't forget to include maintenance.

Model assignment

Scenario

You have been asked to research a range of different bike lights, like the one shown in Figure 3.18, and to investigate how one has been designed and manufactured through performing safe disassembly.

Your task

Perform the safe disassembly of a bike light. The disassembly must be structured and documented, and you must consider and address working safely.

Through disassembly, undertake an analysis of the bike light to include components, assembly methods, materials, production methods and manufacturing processes and maintenance.

Figure 3.18 A bike light

Example candidate response

Plan and Instructions for Disassembly

Tools and equipment	Procedure	Health and safety	Special instructions	Issues/Comments
Latex free gloves	Looking directly at the lens of the bike light, twist the lens cover clockwise to unlock. Remove the lens cover and place in plastic tray labelled 'external body parts'.	Latex free gloves protect hands from lubricants etc. that may irritate skin. Wear PPE.	When removing lens cover, apply a small pressure to the lens cover to prevent damage to the location pins that lock the lens cover to the main body.	Issue – I had to place the product in a vice to gain extra strength when removing the lens cover as it was stiff.
Latex free gloves	Remove the internal components of the bike light from the main body. Place the main body of the bike light into the plastic tray labelled 'external body parts'. Remove the plastic reflector for the light emitting diodes and place in the tray labelled 'internal components'.	Latex free gloves protect hands from lubricants etc. that may irritate skin. Wear PPE.	Lens cover is held in place by an insertion fit with the LEDs and the lens cover.	Comment – If the batteries were still in, and had been used recently, there may have been a slight charge in the capacitor which could have resulted in a light shock.
Pozi-drive screw driver	Looking at the LEDs, identify the M2 pozi-drive self-tapping screws. By using a pozi-drive screw driver, twist the screw driver and the self-tapping screw anti-clockwise to remove. Place screws into the tray labelled 'internal components'.	When handling self-tapping screws, beware of the sharp point of the screw and thread as there is a risk of cutting skin. Return screw driver to the tool case to keep work area tidy and safe. Wear PPE.	Self-tapping screws require very little force to remove. Excessive force may damage other internal components.	Comment – Beware of the sharp point when handling the self-tapping screws and thread, as there is a risk of cutting the skin.

Plan and Instructions for Disassembly (Continued)

Tools and equipment	Procedure	Health and safety	Special instructions	Issues/Comments
Latex free gloves and long nose pliers	Remove positive and negative contacts from the battery holder. Bend the nickel plated contacts away from the battery holder and pull away from the location slots. Remove circuit board, tactile switch and battery connections form the ABS coupling and battery holder. Place coupling and battery holder into the internal component tray.	When handling battery contacts, be aware of sharp edges as they could cut skin.	Support nickel coated steel battery contacts when bending to remove from battery holder as they could snap.	Comment – Be careful not to snap the nickel coated steel battery contacts when bending, as the sharp edges could cut skin.
Soldering iron, de -soldering tool and helping hands	Secure circuit board and LED holder into the help hand vice. De-solder the LEDs from the circuit board and the multi-core wires from the circuit board. Place LEDs, tactile switch and battery connection into the internal component tray.	Make sure extraction is on when de-soldering to remove solder fumes. Risk of burning due to the tip of the soldering iron.	Too much heat applied to the LEDs could lead to permanent damage of the LEDs.	Comment – If the batteries were still in, and had been used recently, there may have been a slight charge in the capacitor which could have resulted in a light shock.

The candidate has written a procedure for the safe disassembly of the bike light.

They have also considered tools required and some control measures.

Risk Assessment of Disassembly

Risk Assessment	Likelihood				
Severity	Frequent	Likely	Occasional	Seldom	Unlikely
Catastrophic	Extremely high risk				
Critical		High			
Moderate			Medium		
Not serious				Low	

Describe the hazard (what can cause harm)	Describe the risk (Low/Medium/High/Extremely High)	Identify three control measures	Who is responsible?
You could cut yourself from trying to pry open plastic casing with a flat head screw driver.	This would be medium risk because if I was doing this ,a cut would be likely to happen, and if it did, the injury would be moderate.	1. Read disassembly instructions 2. Remove all screws from casing first 3. Hold work piece in a vice 4. Wear gloves	The user
A cut from unscrewing a screw with a screwdriver as it could slip and cut you.	This would be a low risk because it is unlikely to happen and the severity of the cut would be moderate at most.	1. Read disassembly instructions 2. Hold work piece in a vice 3. Wear gloves	The user
The force from prising apart certain pieces could result in scratches and cuts from the material.	This would be a low risk because whilst the likelihood is occasional, the injury wouldn't be serious at all.	1. Read disassembly instructions 2. Wear goggles and gloves 3. Hold work piece in a vice	The user
You could accidently break or crack the plastic casing when prising it apart, and it could cut someone and hit them in the eye.	This would be a medium risk because it happens occasionally and the severity of the incident could be critical and someone could lose their eyesight.	1. Wear goggles and gloves 2. If you see it starting to crack, try and find an alternative method of taking it apart 3. Keep people a safe distance away	The user
Your skin could be caught between different components when unscrewing or screwing them.	This would be a low risk because although it is likely to happen, the severity isn't serious at all.	1. Read disassembly instructions 2. Hold work piece in a vice 3. Wear gloves	The user

The candidate has produced a risk assessment for the disassembly and has considered relevant hazards and their associated risks.

The risks have been assessed using a table and they have been colour-coded accordingly.

The candidate has identified sensible control measures as well as who is responsible for making sure these are implemented.

Product Disassembly – Photographic Evidence and Annotation

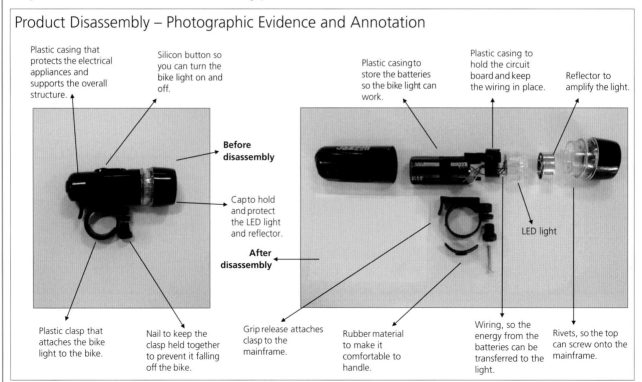

Plastic casing that protects the electrical appliances and supports the overall structure.

Silicon button so you can turn the bike light on and off.

Plastic casing to store the batteries so the bike light can work.

Plastic casing to hold the circuit board and keep the wiring in place.

Reflector to amplify the light.

Before disassembly

Cap to hold and protect the LED light and reflector.

After disassembly

LED light

Plastic clasp that attaches the bike light to the bike.

Nail to keep the clasp held together to prevent it falling off the bike.

Grip release attaches clasp to the mainframe.

Rubber material to make it comfortable to handle.

Wiring, so the energy from the batteries can be transferred to the light.

Rivets, so the top can screw onto the mainframe.

Annotated photographic evidence is used to show the component parts of the bike light after disassembly. This includes details of the components themselves, materials and assembly methods.

There are no annotated photographs showing step-by-step disassembly taking place, including the safe use of tools and control measures.

Analyse Product after Disassembly

Function:
The function of this part is to attach the bike light to the actual bike. It is removable (so it can be removed from the mainframe of the bike light) but also very flexible in terms of where you can place it on your bike. The nail ensures that the bike light stays in place, but it is also removable and temporary, meaning the biker can move the bike light to different places. The clasp also acts as a handle for the user when they are having to carry and transport the bike light to various different places. The curved structure makes it easy to hold.

Manufacturing methods:
The majority of this part was made using injection moulding because it allows the designers and manufacturers to produce shapes that maintain a much more curved shape and still achieve vast amounts of dimensional detail.

Assembly methods:
This handle can be attached to the mainframe of the bike by a slot that has been made in the main body of the light. It is easily detachable making it very versatile. The screw is also easily removable to make sure the bike light can be easily removed and transported elsewhere.

Name of part:
This part is called the handle for when the user needs to carry/transport this bike light. It is also called a clasp because it attaches the bike light to the bike.

Improvements/issues:
This piece could be improved by a thicker material at the bottom of the curve. The constant removal of this clasp must cause the plastic to weaken over time. Extra thickness would ensure sustainability and durability.

Materials
This part of the bike light is made from plastic that has been shaped into a curved structure. The use of this material was employed because it is a strong, durable material that doesn't easily break or crack. It can withstand the constant and relentless exposure to different weather environments. Since the clasp also boasts a very complex structure, plastic is a very flexible material that can create very detailed and delicate forms. Plastic is also a very cheap material to produce and manufacture. The additional rubber material at the base of the curve also presents extra comfort when handling and carrying the bike light. It also provides extra grip for the bike light and adds extra stability when attached to the bike.

Maintenance:
This part can be simply cleaned using some water and soap. If you lose or misplace the nail, it is easy to re-purchase and replace. However, the plastic structure part is not so easily fixable. Whilst the chances are unlikely, if the plastic snaps, the result is irreversible in the fact that you won't be able to fix it. Since it's a very thin structure, the likelihood is more probable than that if it were a thicker material.

Analyse Product after Disassembly

Function:
The function of this plastic casing is to cover and protect the LED light and the reflector. It prevents the LED from getting damaged when dropped and protects it from the constant exposure to varied weather such as rain and snow. The purpose of the light reflector inside the cap is to amplify the light. This ensures full visibility and awareness of the user, which helps avoid accidents and injuries.

Manufacturing methods:
This part was presumably made using injection moulding. This particular process can create smoothed, curved surface, but can also produce finer, more defined details. The rivets in the clear plastic, that allow the cap to be screwed on, were made by injection moulding because that particular process can delicately manipulate plastic to create small but necessary details.

Assembly methods:
This part can be attached and detached to the mainframe of the light by simply screwing it on, twisting clockwise for release and anti-clockwise to tighten. The light reflector and LED light fit easily inside the cap for a compact, easy assembly.

Name of part:
This part of the bike light is normally referred to as the cap.

Improvements/issues:
I think this part of the bike light could be improved by added structural support inside the cap to hold the LED light and reflector in place. This will steady the beam of light and prevent minor damages from slight movement.

Materials:
This part of the bike light is made from acrylic that has been shaped in a curved, aerodynamic structure. The use of this particular material was employed because it is a strong, durable material that doesn't easily break or crack. It can also withstand the constant and relentless exposure to different weather environments. This trait is obligatory for this particular component, considering it has to protect the LED light and reflector inside. Acrylic is also a very transparent material that manages to let 92% of light pass through. Acrylic is a thermoplastic, meaning it can be heated up and moulded into different structures for a different use, which helps the environment.

Maintenance:
This plastic structure feature is not easily fixable. Whilst the chances are unlikely, if the plastic snaps into separate pieces, the result is irreversible. But, if it's just a small crack, it might be temporarily fixable by using duct tape or using super glue. This solution is temporary, however, and in the end, you will probably require a new bike light. If the LED light, the rivets on the clear plastic or the reflector breaks, a new bike light will be required.

The candidate has analysed in detail each component from the bike light, including its function, manufacturing and assembly methods, materials and maintenance. They have also made suggestions for how the part could be improved.

Topic area 2 Modelling design ideas

2.1 Methods of modelling

Virtual CAD 3D modelling

As we saw in Unit R039, a popular method for communicating design ideas is using virtual CAD models. These models can be used to show concept designs to clients and customers for their feedback and are much quicker to produce and modify than creating physical prototypes. However, they can also be used as the starting point for later creating physical prototypes by hand or using computerised methods, such as rapid prototyping and laser cutting.

CAD software allows virtual models to be produced by creating parts using sketch-based tools and building these into assemblies using mates and constraints. These virtual models, as we have already seen, can be suitably rendered to make them look realistic. They can also be animated to produce video demonstrations showing the product, and even its component parts, moving and rotating.

Simulating the operation and interaction of connected parts within an assembly is relatively easy with CAD software. The software also often allows much more complex simulation, as already highlighted in Unit R039.

A popular CAD simulation tool is a **motion study** (sometimes called a movement study). This allows designers to study the motion of different components under different input conditions, such as pushing and pulling or rotating a component.

Key term

Motion study A simulation tool in CAD software allowing the study of how parts and components move in relation to each other.

A simple motion study also allows investigation of how well components fit together and to check if they fit together easily, are a tight fit or will not fit together at all. It can also be used to check the tolerances are correct between components.

More complex motion studies can even simulate the effects of things like gravity on components, can be used to simulate the effect of springs and can be used to check that moving components do not collide or crash together (if they are not meant to) when moved.

Here is a simple motion study for a virtual model of a glue gun. For the purposes of producing a virtual model, the glue gun only has six main components: body, trigger, heater, glue stick push lever mechanism, spring and nozzle. The model has been created using the CAD drawing techniques already covered in Unit R039 (page 143) in preparation for simulation.

The virtual model, shown in Figure 3.19, can be simulated to confirm the correct assembly and operation of the glue gun. The simulation confirms that:

- The trigger rotates around a point when pressed and is pulled back by the spring when released.
- The trigger is correctly linked to the glue stick push lever.
- The glue stick push lever slides forwards and backwards as the trigger is pressed and released – which moves the glue stick forward.

The simulation shows both **rotational movement** and **linear movement** when the trigger is pressed and released. The complete glue gun assembly is also shown.

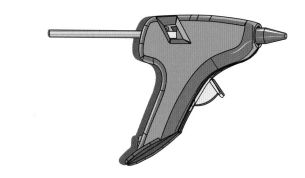

Glue gun assembly

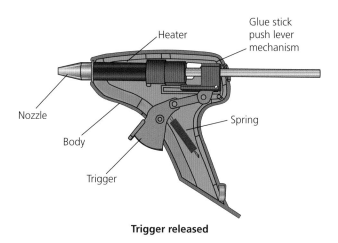

Trigger released

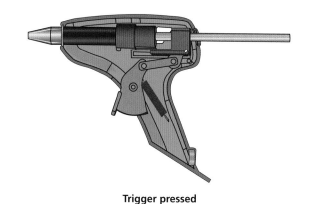

Trigger pressed

Figure 3.19 Glue gun motion simulation

Key terms

Rotational movement Movement around an axis, such as hinge movement.

Linear movement Movement in a straight line, forwards and backwards.

Activity

Perform a simulation using the CAD model of the roller assembly that you created in Unit R039. Use your model to show how the roller can be rotated within the supporting brackets. You might need to put a mark on the roller to show clearly that it is being rotated!

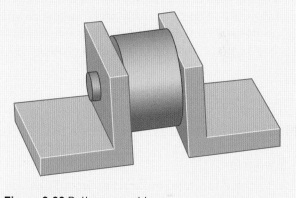

Figure 3.20 Roller assembly

Research

Use the internet to find videos and tutorials of how to use CAD software to set up and perform a motion simulation. You will find many examples by searching the internet and YouTube for 'CAD motion simulation'.

Test your knowledge

1 What are the advantages of using virtual 3D modelling and simulation?
2 How are virtual models created using CAD software?
3 What is a motion study, and how can it be used to investigate how the virtual prototype operates?

Assignment practice

Marking criteria

Mark band 1: 1–4 marks	Mark band 2: 5–8 marks	Mark band 3: 9–12 marks
Produces a **basic** 3D virtual model using CAD.	Produces an **adequate** 3D virtual model using CAD.	Produces a **comprehensive** 3D virtual model using CAD.
A **simple** 3D virtual model consisting of a very limited number of components.	An **adequate** 3D virtual model consisting of some mated components.	A **complex** 3D virtual model consisting of many mated components.
Production of a 3D virtual model is **dependent** upon assistance or help from other sources.	Production of a 3D virtual model is carried out with **some** assistance or help from other sources.	Production of a 3D virtual model is carried out **independently**.

Top tips

- For this unit, you will be given a set assignment brief containing a scenario and tasks. Read this carefully and make sure you address all the points in the marking criteria. The scenario will be based on a product.

- For this part of the assignment, use your skills at using CAD software to model the design provided.
- Show how movement in your CAD model can be simulated.
- Use step-by-step annotated screenshots to communicate how you have created your CAD model and to show how the simulation works.

Model assignment

Scenario

A portable lamp manufacturer would like you to create a 3D model based upon their own product design and specification. The drawing is shown below.

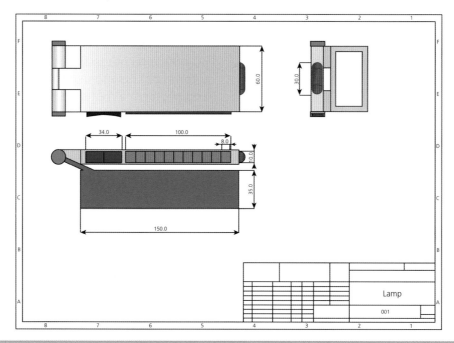

Your tasks

● Use CAD software to recreate an accurate virtual 3D model of the portable lamp, producing individual components and using the mate tool to create an assembly.

● Use CAD software to simulate the operation of the portable lamp.

Example candidate response

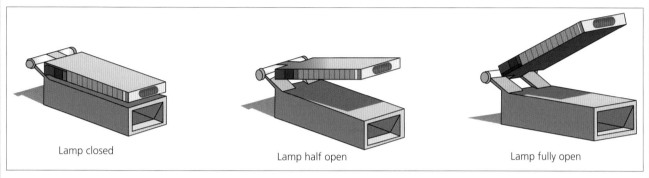

Lamp closed   Lamp half open   Lamp fully open

The candidate has taken the information on the supplied orthographic drawing and used this to create individual components and assemble these using the mate tools to produce a 3D isometric model.

The software has been used to simulate motion of the lamp being opened and closed, with the three images showing the lamp closed, half open and fully open.

Physical modelling

Getting started

Consider a simple piece of packaging, such as a box. It is a product that will have been designed to contain and protect what is inside it. A box is constructed with folds, joining seams, flaps and closures, and may also have additional components to further secure its contents in compartments. The telecommunications giant Apple boxes its products in highly developed packages with attention to detail that suggests high quality even before you get to the product

in the package. The company has taken what was a simple piece of packaging and turned it into an important component of the purchase experience.

Produce a quick pre-prototype sketch-model piece of packaging for a small product such as a pencil or pen. Describe how the package could be designed to suggest quality and add to the purchase experience.

The desired outcome of physical modelling is a 3D prototype model; achieving a successful outcome requires a systematic approach that includes planning, production and evaluation.

Modelling, often called prototyping, is a process used prior to manufacturing products to prove

that theories or designs actually work as they are supposed to do. The materials, tools, equipment and processes you select to make a prototype model will depend on the purpose for the model – will it be a **functional prototype** model or a **concept prototype** model?

A functional model is a working model; it appears and performs as a final product might and can be used to test performance or endurance. Prior to production of products, manufacturers and consumers need to know that the products will do the job they have been designed to do, and do it safely.

For example, for each new Formula 1 season, racing teams produce new cars with the hope of them outperforming the cars of other teams. In the development stages, functional prototype racing cars are produced, which can be driven and tested. In the final stages of prototype development, prototypes will often be manufactured using the materials and components planned for the final product.

Figure 3.21 A functional prototype Formula 1 racing car

might, in terms of style, shape, form, texture and colour, and are often produced to test market reaction or for a pre-production sales promotion. Concept models may be larger or smaller than the product they have been designed to characterise.

Concept models can be made to a very high quality, so much so that it might be almost impossible to tell the difference between the concept model and the fully functioning product from photographs. The concept model of a drone shown in Figure 3.22 appears to have all of the components required for it to take off and fly, but it is a non-functioning model with no internal components.

Figure 3.22 A concept prototype model of a drone

Once the engineer has a clear understanding of the specification requirements and a decision has been made about the type of prototype to be developed, the next stage is to identify the general order of the processes for making a prototype model.

This is an opportunity to review your planning and make it more logical. This can be shown in a flowchart, which is a diagram that shows how individual processes are interlinked, such as Figure 3.23. The diagram does not need to communicate every separate activity within each process. The different symbols represent stages within the overall process and show how the various processes are linked.

Key terms

Functional prototype Prototype model that has the appearance and performance of a final product.

Concept prototype Prototype model that appears as a final product might but often does not function as a fully working product would.

Visualisation Representation of an object.

The materials chosen for concept modelling do not need to be the same as they would be for a finished product because they don't have to perform like a finished product. Concept models are aesthetic **visualisations**; they appear as a final product

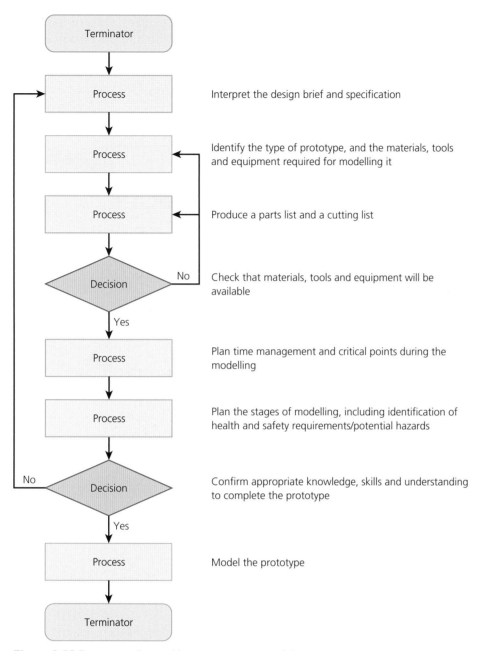

Figure 3.23 Processes for making a prototype model

Activity

In pairs, review the strengths and weaknesses of functional prototypes and concept prototypes. Give examples of products that might require each type of prototype.

When might concept prototype modelling not be an appropriate process?

Production planning

Planning involves identifying and communicating the activities that are required to achieve a desired outcome. Effective planning is essential for efficient use of time and resources, including money. Initially, it is a thought process where all the stages and processes are scheduled in an effective order.

When planning the making of a prototype, a range of planning tools, such as flowcharts, Gantt charts, task lists and planning tables, are available to help with different aspects of the planning. For example, they can identify what should be happening when and where, with which materials, tools and equipment, and by what point in time. Planning documents are often linked to a range of other documents, such as health and safety guidance, risk assessments, information about personal protective equipment (PPE), legislation and quality standards.

Engineering drawings provide the starting point for planning documents – it is from these that all of the parts or components that are to be manufactured can be identified and **cutting lists** can be produced.

Key terms

Engineering drawing Type of technical drawing that details the geometry, dimensions and features of a component or product.

Cutting list List of the materials required at their prefabrication size so that modelling or making can commence.

Parts list List of all of the individual components to make a working product, which usually includes component names, quantities, materials, costs and suppliers.

Exploded assembly A drawing where the components of a product are drawn slightly separated from each other and suspended in space to show their relationship or the order of assembly; also known as an exploded view.

Parts list

A **parts list** will have an image of the product to be manufactured presented as an **exploded assembly** drawing, with all of the individual components required to make a working product. Each of the parts will be labelled and named, as

shown in Figure 3.24, or given an item number and tabulated by description and part number, as shown in Figure 3.25.

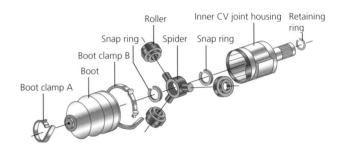

Figure 3.24 A parts list with components named

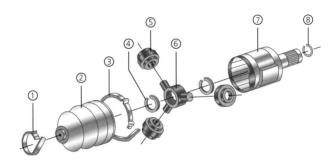

Number	Description
1	Boot clamp A
2	Boot
3	Boot clamp B
4	Snap ring
5	Roller
6	Spider
7	Inner CV joint housing
8	Retaining ring

Figure 3.25 A parts list with components numbered and listed

Cutting list

A cutting list provides information about the components to be produced, by indicating the materials required before modelling or making commences. It is generally presented in a table, identifying the materials and their general dimensions, the quantity required and any important notes.

Table 3.15 An example of a cutting list for a craft-modelled computer mouse

Component number	Component name	Component description	Material	Quantity required	Length (mm)	Width (mm)	Height (mm)	Thickness (mm)	Modelling production method	Notes
001	Lower casing	Lower casing for battery, electronics and optical reader	Closed-cell extruded polystyrene foam	1	100	55	25	Solid block	Craft-based fabrication	HD foam, shape to required dimensions, smooth finish
002	Upper casing	Upper casing	Closed-cell extruded polystyrene foam	1	10	55	20	Solid block	Craft-based fabrication	HD foam, shape to required dimensions, smooth finish
003	Battery closure	Lower casing battery enclosure	HIPS	1	50	45	5	1	Craft-based fabrication	
002	Left-hand switch button	Left-hand switch button	HIPS	1	40	25	2	1	Craft-based fabrication	Thickness is nominal
005	Right-hand switch button	Right-hand switch button	HIPS	1	40	25	2	1	Craft-based fabrication	Thickness is nominal
006	Roller switch button	Roller switch button	ABS	1	20 diameter			5	Craft-based fabrication	

Flowcharts

A **flowchart** is a block diagram that shows how various processes are linked together to achieve a specific outcome. Each of the main processes is identified and checkpoints (questions/decisions) are added for clarification. Where a question or decision cannot be answered with a positive response, there is a return (feedback) to an earlier process or to the beginning of the flowchart.

The example in Figure 3.27 shows the processes for drilling a centre hole using a pillar drill. Feedback loops are shown where a 'no' response is possible and it is necessary to return to an earlier process.

A flowchart can be presented as a simple diagram, with each process block representing several processes, or they can be very detailed, with multiple options and links to further flowcharts.

Key term

Flowchart Block diagram that shows how various processes are linked together to achieve a specific outcome.

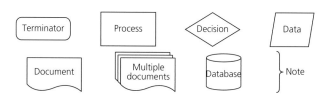

Figure 3.26 Common flowchart symbols

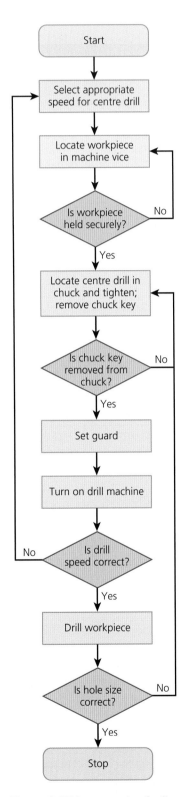

Figure 3.27 An example of a flowchart for drilling a centre hole using a pillar drill

Gantt charts

The **Gantt chart** is named after Henry Gantt, who developed the model in around 1917. It is still a widely used project-planning tool. It comprises a sequence of activities or operations listed vertically with corresponding timescales plotted horizontally.

When planning timescales for each activity, it is important to be realistic, taking into account access to the facilities you will require (for example, a certain machine that other people might need to use). Not all projects are linear (with one activity occurring after another), so it is possible to have some processes planned as parallel activities (more than one activity happening at the same time) when you may be waiting for a tool or machine. It is also useful to add some important checkpoints or project markers to indicate where reviews are due.

Activity time for short projects (3–5 hours), such as modelling a prototype, should be manageable, timed sessions of about 15–30 minutes each, rather than week by week, so, for instance, a one-hour lesson can be broken down into four 15-minute sessions, as in Figure 3.28. In this example, linear activity is indicated by a blue line, checkpoints by a red diamond and evaluation by a green line.

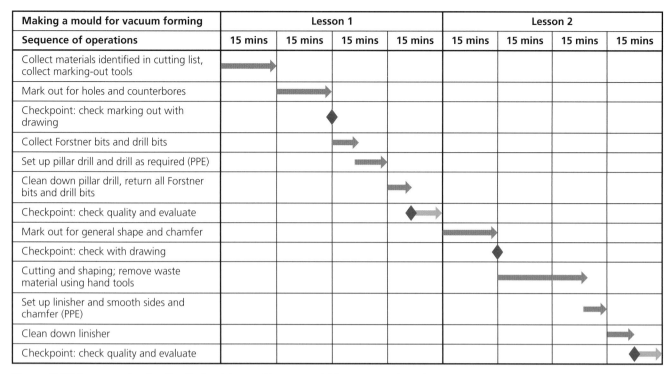

Making a mould for vacuum forming	Lesson 1				Lesson 2			
Sequence of operations	15 mins	15 mins	15 mins	15 mins	15 mins	15 mins	15 mins	15 mins
Collect materials identified in cutting list, collect marking-out tools								
Mark out for holes and counterbores								
Checkpoint: check marking out with drawing								
Collect Forstner bits and drill bits								
Set up pillar drill and drill as required (PPE)								
Clean down pillar drill, return all Forstner bits and drill bits								
Checkpoint: check quality and evaluate								
Mark out for general shape and chamfer								
Checkpoint: check with drawing								
Cutting and shaping; remove waste material using hand tools								
Set up linisher and smooth sides and chamfer (PPE)								
Clean down linisher								
Checkpoint: check quality and evaluate								

Figure 3.28 An example of a Gantt chart for making a mould for vacuum forming

Key term

Planning table Formal planning sheet that records all the relevant information required for modelling and making.

Planning tables

A **planning table** is a method of formally recording all the relevant information required for modelling and making. Each stage for each individual component should be identified and described in such detail that someone else could work from it without needing clarification.

Table 3.16 is an example of a planning table and includes:

- the steps (or stages)
- the processes, materials, tools and equipment for each stage
- relevant health and safety requirements
- relevant quality control management
- time allocated
- space for additional information.

Table 3.16 An example of a planning table

Step	Description	Material(s)	Tools/ equipment	Health and safety/PPE	Quality control management	Time (min)	Additional information
100							
120							
130							
140							

Detailed plan development requires the person who writes up the plan to have a clear understanding of what is to be made (each part or assembly) and how it is to be made (for example, the resources available to make it). When producing the planning table, it is useful to number the stages in such a way that extra stages can be inserted later without disrupting the whole plan. For example, extra stages after Step 120 could be numbered 121, 122 and so on.

The final column, 'Additional information', can be very useful for adding details that could help someone else to use the table – for example, alternative choices or possible changes that may be necessary.

Planning for CNC manufacture

For modelling and making using a machine linked to a computer (for example, a laser cutter, CNC milling machine, lathe, router or 3D printer), the planning document can be quite similar to that used for craft-based modelling, with each stage identified along with the process and equipment.

The starting point of the plan is creating the 2D drawing or 3D model onscreen using CAD software and then saving it with a file extension that is compatible with the CNC machine, such as .stl, .fnc, .dwg or .dxf. The conversion to these file extensions changes the screen image format to a numerical code such as G-code, the programming language of coordinates for computerised automated manufacture.

Once the converted file has been uploaded to the computer, the CNC machine settings can be adjusted and a simulation should be run to check that successful machining could be achieved. It is helpful to add stages into the plan such as machine setup and clean down after use, along with all appropriate tools, equipment, PPE and health and safety considerations.

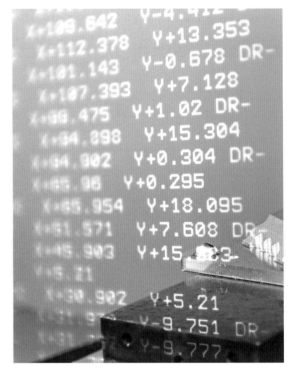

Figure 3.29 An example of G-code

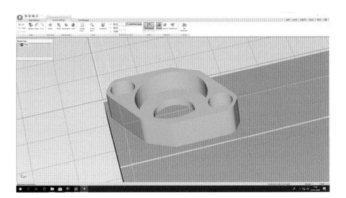

Figure 3.30 An example of a 3D virtual model created using computer software

Figure 3.31 Production of a 3D model by rapid prototyping

Planning resources

To be able to add specific features to your plan for making a prototype, you must have clear understanding of what is going to be made and the resources available. It is adding this level of detail that makes the plan useable for other people.

Resources include:

- the materials that the prototype will be made from, such as **polymers**, woods, boards and metals
- the tools/equipment required to make each part of the prototype.

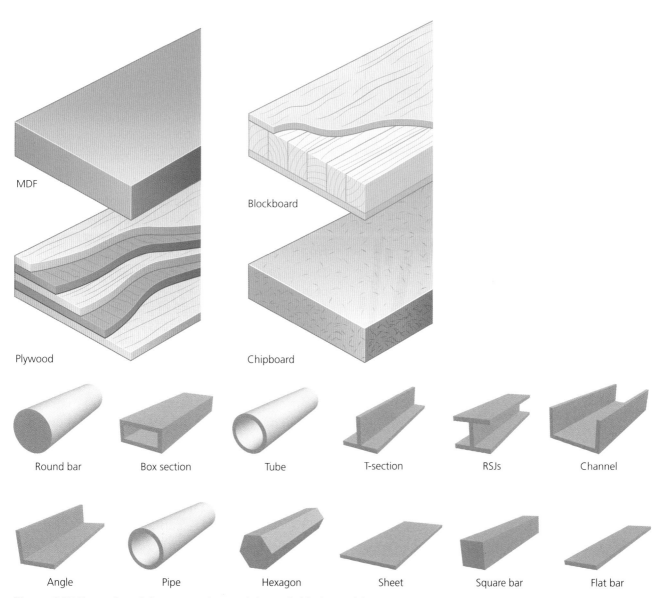

Figure 3.32 Examples of the range of materials available for making prototypes

The materials chosen for the prototype will depend on the type of prototype to be made (concept or functional). The prototype may also be made up of several parts or assemblies that are each made from different types of material; the planning and making of each part is equally important.

Each part may include standard components that can be bought in, such as nuts, bolts, springs and washers, or components may be made specially if they are specific to the prototype.

a)

b)

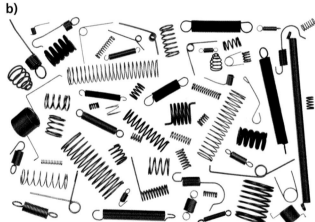

Figure 3.33 Examples of 'off the shelf', bought-in components

Activity
Analyse the merits of using standard and/or pre-manufactured components when manufacturing.

The tools and equipment used will also depend on the type of prototype.

For craft-based modelling of a concept prototype, the following tools and equipment could be used:

- for marking out – pencil, marker pen, stencil, flexible curve, steel rule, engineer's square, divider, odd-leg calipers
- for wasting – modelling knife, scalpel knife, hand saw, scroll saw, band saw, circular saw, linisher, bobbin sander, router, wood lathe, hand drill, pillar drill
- for finishing – glass paper, wet and dry paper, paint, oil, varnish, wax
- for assembly – adhesive, permanent fixings, non-permanent fixings.

For a functional prototype, the following tools and equipment could be used:

- for wasting – centre lathe, milling machine, grinding machine, CNC machine
- for accurate quality control – micrometer, depth gauge, vernier calipers, vernier height gauge.

Once the materials have been identified, a parts list and a cutting list should be produced.

Health and safety in planning

The recognition of potential hazards and risk of injury is an important aspect of planning. Therefore, at each stage of planning, when the tools, equipment and processes are considered, consideration should also be given to safe working practices.

The engineering workshop presents many different potential hazards, in terms of materials, tools, equipment and processes. Legislation such as The Health and Safety at Work Act 1974 (HASAWA 1974) and Health and Safety Executive (HSE) regulations require workers to work safely and not to put themselves (or those around them) in danger of harm or injury where it can be avoided. In order to comply, engineers are required to follow safe working procedures. When tools and machinery are used safely and correctly, the risk of injury is low.

The production plan does not need to list all the hazards, risks and control measures, as that is a purpose of the risk assessment, but it should show that they are recognised and have been considered; the plan should identify the relevant personal protective equipment (PPE) for each stage of prototype modelling.

An example of this is when someone uses a sanding machine for concept modelling. The sanding machine is a potential hazard and the user must be aware of how to be safe when operating it. In the production plan, the sanding machine should be named, with a reference linking it to the relevant risk assessment for **abrasion** tools, including sanding machines. In terms of PPE, a face mask or goggles will be required, along with an apron or work coat to hold back any loose clothing. Appropriate **local exhaust ventilation (LEV)** for dust extraction may also be required and, if so, this would also be added to the production plan.

Key terms

Abrasion Process of wearing something away.

Local exhaust ventilation (LEV) Control system designed to reduce exposure to airborne pollutants and contaminants, such as dust, fumes, gas and vapour, by taking them out of the workplace.

Cutting and wasting Removal of unwanted material.

Table 3.17 An example of one stage in a production plan, including recognition of a potential hazard, risk assessment, relevant PPE and LEV

Step	Description	Materials	Tools/ equipment	Health and safety/PPE	Quality control management	Time (min)
150	Remove excess material by abrasion to form the general shape of the hand controller	Jelutong	Linisher sanding machine	Hazard: sanding machine (see Risk Assessment DT2019:105 – Abrasion tools) PPE: face mask/goggles; apron/work coat LEV is required	Drawing; template; radius gauge; steel rule; vernier caliper	15

In order to accurately plan for safe working when prototype modelling, the engineer must have an excellent understanding of all relevant safety considerations. This can be achieved by:

- being aware of the hazards presented by the tools/machines to be used and reviewing all appropriate risk assessments
- being aware of the hazards presented by the location (the place where the tool/machine will be used); this can usually be done by visually checking the work area
- fully understanding how to use the planned tools/machines safely and be aware of any special instructions for their use.

Tools and processes with a particularly significant risk of injury when modelling include:

- tools for **cutting and wasting** (for example, sharp-edged tools, such as saws and drills); however, all tools should be used with care because each presents a potential risk of injury
- processes that produce airborne particles (for example, using a sanding machine)
- processes that generate heat (for example, using a strip heater, oven, vacuum-forming machine, laser cutter or 3D printer).

The engineer should ask themselves a series of questions before planning and starting any practical work, as outlined in Table 3.18.

Table 3.18 Questions that should be asked before starting practical work

Question	How to answer
What tool/equipment/process is to be used in this stage of making?	Look closely at the production plan and at the relevant stage.
What tool/equipment/process is to be used in the location by people around me?	Pay attention, use your observation skills and talk with others to check what you need to watch out for.
Is there a relevant risk assessment for the tool/equipment/process, and has it been localised specifically for this workshop?	Review the risk assessment and check that the particular tool/equipment/process is identified. For tools and equipment, the serial number should be listed.
What are the risks and who is affected?	Read the risk assessment carefully and make a note of the risks and to whom they apply.
What are the control measures?	Read the risk assessment carefully and make a note of the control measures.
What PPE is required?	Read the risk assessment carefully and make a note of the PPE that will be required.
Have I been appropriately trained to use the tool/equipment/process and am I confident to work with it?	The engineer must be honest with themselves. Working with a tool, piece of equipment or process that you have not been appropriately trained to use or are not confident in using is one more hazard to the workshop. You must not work in such a way that you yourself present a risk to others.
Is the tool/equipment safe to use?	Check the tool/equipment to make sure it is in such a condition that it is safe to use; it must not be overly worn or show any signs of damage. Excessive wear or signs of damage should be reported and the tool/equipment must not be used.

Activity

Produce a flowchart that presents actions (processes) and questions that should be considered for safe working prior to starting work using a tool, piece of equipment or process.

While most tools, equipment and processes present some risk of injury or damage, there are some that require extra vigilance and care, such as sharp edge tools, and tools for cutting and shaping.

Modelling knives are an example of sharp-edged tools. The blade should only be exposed when the knife is to be used; when not in use, the knife should be located in a safe place, such as away from the edge of a bench, and not carried around in an apron or a trouser pocket, unless they have added protection for the blade area.

When using shape edge tools, such as a modelling knife, you should never cut towards yourself, and where possible you should use a safety rule or safe edges.

Hand saws are tools that have angled metal teeth. Control the cutting stroke carefully, where possible by using two hands. For saws where only one hand is used to grip the handle, the other hand should be kept well clear of the cutting area.

Figure 3.34 An example of safe working: the engineer is aware of potential hazards around his work area and is wearing appropriate PPE

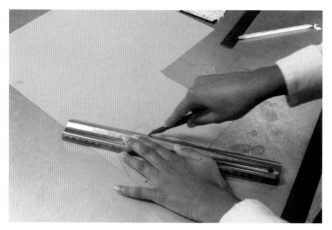

Figure 3.35 Safe working when using a modelling knife

Figure 3.36 Safe working when using a hacksaw

Sharp-edged tools should:

- be stored in such a way that they do not present a risk of injury when selected
- only be used after appropriate training, under general supervision and for the purpose for which they were designed
- be maintained such that they remain fit for purpose.

Key terms

Portable power tools Motor-driven tools that can be carried around and used in different locations – for example, a hand drill or a palm sanding machine powered by battery or by mains electricity.

Fixed-position power tools Motor-driven tools that cannot practically be moved and are usually bolted down – for example, a pillar drill.

When using **portable** or **fixed-position power tools**, basic safe working practices include:

- When collecting and returning battery-powered hand tools, where possible, the trigger should always be in the locked-off position.
- If a battery becomes hot during use or starts to give off smoke or fumes, the tool should not be used and it must be reported immediately.
- Where a power tool has a power cable, pay attention to its location.
- Disconnect portable mains-powered appliances from the power supply when not in use.
- Fixed-position tools should always have their power supply turned off and isolated until they are required.
- Appropriate PPE must always be worn when using power tools, both by the user and by those nearby.

Specific safe working practices depend upon the tool being used.

When using a jigsaw, it is important to securely hold the material being cut, as shown in Figure 3.37. The user should have one hand on the handle of the jigsaw to press the trigger while the other hand lightly presses the tool down and guides it as it cuts, the blade moving in an up-and-down stroke.

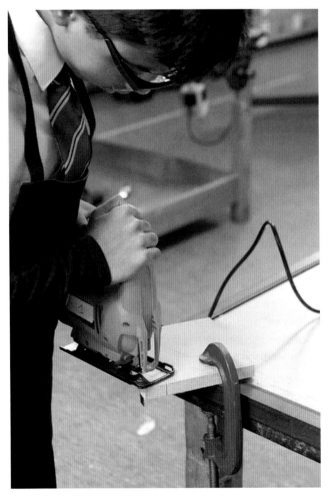

Figure 3.37 Safe working when using a jigsaw

Figure 3.38 Safe working when using a powered hand drill

When using a powered hand drill, the material to be drilled should be placed on a **sacrificial board** above a solid work surface and securely held using an appropriate clamping device, as shown in Figure 3.38. The location of the hole should be clearly marked beforehand and initially drilled with a centre drill or small pilot-hole drill. One hand should be on the handle controlling the on/off trigger, and the other should be guiding the drill to the correct location.

Key term

Sacrificial board A flat piece of board that can be drilled and damaged.

When using a pillar drill, the workpiece should be securely held in a hand vice or fastened to the table, and all guards must be closed to prevent debris flying out and towards the operator. All drilling should be performed below eye level, as shown in Figure 3.39.

Fixed-position sanding machines are extremely good at removing excess waste due to the abrasive material on their disc or belt. Safe working practices include limiting the exposed abrasive surface to what is required; the rest of the abrasive surface should be guarded, as shown in Figure 3.40. The work should be securely held by the engineer's fingers, which should be located well away from possible contact with the abrasive surface.

Figure 3.39 Safe working when using a pillar drill

Figure 3.40 Safe working when using a belt sanding machine

Processes that produce airborne particles

Not every potential hazard is clearly visible. For example, airborne particles – though they are not always a risk to health and safety, every effort should be made to minimise them.

Abrasive processes, such as sanding, produce airborne particles. These tend to be quite fine and lightweight. Timber and manufactured-board dust can be particularly irritating if it becomes airborne, and therefore should be removed by LEV.

However, small hand-held sanding machines, such as orbital and detail sanders, are generally used away from LEV systems. When used to sand wood-based materials, they tend to produce fine dust. Their use should be limited to light sanding only, and they should never be used in confined workplaces without adequate ventilation. Engineers should wear appropriate PPE for eye and respiratory protection.

Aerosol spray adhesives, lacquers and paints have a strong, unpleasant odour, and may be highly flammable and dangerous if inhaled. Their use should be carefully monitored by a responsible person and confined to a spray booth or to where there is exposure to a mobile air supply (such as outside areas).

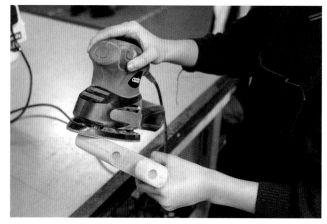

Figure 3.41 A hand-held orbital sanding tool being used for light sanding

Key term

Aerosol Mixture of liquid droplets or minute solids in air or another gas – for example, spray adhesive or spray paint.

Processes that generate heat

Various types of craft-based modelling materials can be shaped and formed using heat. For example, at temperatures around 140 °C, **thermoforming polymers** become pliable so that they can be bent, twisted and folded, and typical processes include oven heating and strip heating. At above 180 °C, many change to a plastic state where they can be moulded and formed; oven heating and vacuum forming are processes often used at the higher temperatures.

Heat can also be used to cut modelling materials. For example, expanded modelling foam melts when pushed against the taut metal wire of a hot wire cutter at a temperature of approximately 200 °C.

A laser cutter is a computer-controlled machine where a laser beam generates intense heat and can be used to cut through a wide range of materials, including woods, polymers and some metals.

Additive-manufacturing machines, such as 3D printers, are also heat-producing tools, but rather than cutting through materials, they deposit molten polymer layer by layer, forming a product on a base plate.

Heat-producing tools should never be left unattended when turned on and should be adequately ventilated. Operators should also wear appropriate PPE.

Stretch activity

Research LEV to identify the type of machines in the workshop for which it is a compulsory requirement. Explain why LEV is a statutory requirement for some machine processes.

Use of PPE during production processes

Wearing PPE is not just something that is good to do, it is a legal requirement for those who work within dangerous environments. The Personal Protective Equipment at Work Regulations 1992 state that where risks cannot be controlled by other means, appropriate PPE *must* be available. Workplaces must provide equipment for users but users must also take reasonable steps to ensure the PPE is used properly.

In workplaces that present specific risks, the engineer should be able to see clear signage that indicates the PPE required. Circular signs with a blue background and white image are called **mandatory signs** and these are used to indicate that an action must be taken – for example, when PPE must be worn.

Key term

Thermoforming polymers Polymers that become pliable when heated, such that they can be shaped and formed, and harden when cooled.

Mandatory signs Safety signs that indicate an action must be taken.

Figure 3.42 Examples of mandatory PPE signs

Other safety signs that may be found in the workshop include:

- prohibition signs that indicate behaviour that is not allowed (white circle with a red outline, with a black image identifying what is prohibited and a red diagonal line through it)
- hazard signs that warn of a specific danger (yellow triangle with a black outline and black lettering/image)
- safe condition signs that show directions to areas of safety and medical assistance (rectangle with a green background and white lettering/image)
- fire safety signs that show the location of fire equipment and compliance with fire precautions (rectangle or square with a red background and white lettering/image).

At times, engineers may be required to wear a combination of PPE, depending on the materials, tools and machines to be used. The basic purpose of PPE is to provide the engineer with protection – for example, from entanglement, impact, soiling, dust, extreme temperatures and high-level noise.

The main types of protection provided by specific items of PPE are:

- body protection (for example, apron, work coat, boiler suit, flameproof overall, weatherproof clothing)
- head protection (for example, safety helmet, cap, cap with snood, welding helmet)
- eye protection (for example, visor, safety glasses, goggles, welding mask)
- ear protection (for example, ear defenders, ear plugs)
- respiratory protection (for example, disposable filter mask, half mask, full-face mask)
- foot protection (for example, safety shoes, safety boots, wellington boots)
- hand protection (for example, barrier cream, disposable work gloves, safety gloves, cut-resistant gloves, heat-resistant gloves).

Figure 3.43 Examples of PPE

Activity

Produce a PowerPoint slide or poster to explain the range of safety signs that can be found in the workshop (mandatory, prohibition, hazard, safe condition and fire safety signs).

Safe working procedures when using materials, chemicals, finishes and solvents

The **Control of Substances Hazardous to Health (COSHH) Regulations 2002** provide a framework to protect people in the workplace against health and safety risks from being exposed to hazardous substances.

An engineer may be required to work with hazardous substances. Manufacturers and suppliers of such substances are required to provide **safety data sheets (SDSs)** for their products, which include information such as potential hazards, how the products should be handled and stored, and emergency measures in case of an accident. All this information is essential for a risk assessment. Engineers should try to select materials, chemicals, finishes and solvents that present the least risk to their health and safety and follow safe working procedures.

Before they are cut, shaped or finished, most modelling materials are relatively safe. However, once the engineer starts to work with them, the procedures and processes could be dangerous.

Key terms

Control of Substances Hazardous to Health (COSHH) Regulations 2002 Legislation that provides a framework to protect people in the workplace against health and safety risks from being exposed to hazardous substances.

Safety data sheets (SDSs) Written documents that provide information and procedures for the safe handling and use of chemicals.

It can be difficult to clean away fine particles of materials dust; damping down dusty areas with water can help to limit the amount of dust that becomes airborne.

Once a timber prototype has been constructed, it may require sealing in order to maintain surface quality or to provide a surface on which a finish can be applied. This involves using products such as polyurethane, shellac and lacquer – sealants that can be oil-based, water-based or a synthetic mix. Use non-toxic and non-flammable finishes if possible.

Polyvinyl acetate (PVA) is often used when joining wood and wood-based materials, but other natural and synthetic adhesives are available; for joining dissimilar materials, a solvent-free contact adhesive may be used, or even a hot-melt adhesive that bonds quickly as it cools. Whichever adhesive is chosen, the engineer should review the appropriate SDS before use and wear appropriate PPE for eye and hand protection.

A wide range of polymers can be used for prototype modelling – for example, nylon, polystyrene (PS) and acrylonitrile butadiene styrene (ABS). These materials can be formed, shaped, cut and machined using a range of processes. Machining processes such as drilling, turning and milling can cause small pieces of polymer debris to fly away from the work area. Therefore, machine guards should be used and eye protection worn in and around the work area. When drilling small polymer workpieces, a hand vice should always be used because a drill bit can bite into the material, causing it to lift and spin with the drill bit.

Thermoforming sheet polymers shaped using heat processes such as oven heating, strip heating and vacuum forming should not be left unattended during heating. This is because they can melt and burn at relatively low temperatures and give off unpleasant and potentially dangerous fumes.

Expanded foam polymers can be rendered and finished to a very high standard, such that concept models can appear as working products. To achieve a high-quality finish, the engineer will often want to seal the surface of the expanded foam prior to rendering it using a colour coat. As with wood-based modelling, when sealants, paints and lacquers are to be used, non-toxic and non-flammable products should be selected where possible.

Aerosol spray paints and finishes have become very popular because high-quality outcomes can be achieved. These present several potential risks to health and safety. Their use should be controlled by following the directions on the can and restricted to well-ventilated areas, ideally in a spray booth. Wherever possible, non-toxic and non-flammable paints and finishes should be used. PPE should be worn in the spraying area, such as full-face masks fitted with appropriate filters.

Activity

List a range of commonly used modelling materials and identify the adhesives, chemicals, finishes and solvents with which they can generally be used.

Figure 3.44 Using a safe procedure when drilling a polymer

Many polymers require the use of a thinner or solvent cleaner before applying an adhesive or joining cement, if they are to be joined to the same material. This is to decontaminate the area of the joint and may be available in liquid or aerosol form. Good ventilation is required and eye protection and gloves need to be worn when applying them. When joining dissimilar polymers, contact adhesives can often be used, but not always, because they can cause some polymers to melt. The SDS for the adhesive will indicate which materials it cannot be used with.

Superglue is widely used for joining polymers and non-polymers. It has a rapid bonding process and can join a wide range of similar and dissimilar materials. Care must be taken not to get superglue on the skin because it will bond almost immediately.

Time requirements

Time is an important resource, and good use of time is highly relevant to planning. Experience of making will help an engineer to estimate how long a task may take, but it is only an estimate. Each stage of making should be given an approximate time period, and the overall time for all stages should match the time given on the Gantt chart; it is easier to manage small blocks of time for an individual activity than for a whole project.

During modelling and making, it is good practice to over-write planning documents where changes occurred – for example, if the time actually taken for an activity was significantly different from the time that was planned for it.

Time management is vital in industry because it is an important factor in calculating the cost of producing products.

Testing and evaluation in planning

Testing and evaluation should not be seen as simply the last thing to do; they are extremely important processes that relate directly to planning and making and should be built into planning as checkpoints, as shown in Figure 3.28 (page 191).

Planned testing, or **quality control testing**, is pre-arranged testing performed in order to produce information about a particular outcome at a particular stage in the process. If the outcome is not as required, evaluation must take place, and a decision must be made about what to do next. Terminal testing is about checking for desired outcomes at the end of a series of processes and is linked to overall **quality assurance** and the monitoring of every stage in the process, from raw materials to production.

At various stages throughout the making process, evaluations or judgements should be made against planning documents and engineering drawings to check that the potential outcome will be fit for purpose. For example, this could be:

- visual checks of shape, fit or finish – does it look as it should or does it fit together as it should?
- dimension checks against drawings and tolerances, using a range of quality-control devices
- functional checks – does the part, component or product do what it is designed to do?

Evaluation is a valuable process if it is reliable – that is, where an evaluation of the same data or product at a different time would produce the same results.

Activity

Take a specification (one you have written or been given) and review the specification criteria. How could you evaluate each criterion (visual, dimensional, function)? Remember, the outcome of the evaluation should be reliable – that is, the same each time, no matter who performs the evaluation.

Test your knowledge

1 How does a flowchart differ from a Gantt chart?
2 What should be included in a planning table?
3 What are some examples of 'off the shelf', bought-in components?
4 How can risk assessments be acknowledged in a planning table?
5 What is the difference between quality control testing and quality assurance?
6 What is the purpose of a risk assessment?
7 What are the main features of a mandatory safety sign?
8 What kind of information will you find on a safety data sheet (SDS)?
9 What are three pieces of personal protective equipment (PPE)? For what are they used?
10 Why should evaluation checkpoints be built into the planning process?

Assignment practice

Marking criteria

Mark band 1: 1–2 marks	Mark band 2: 3–4 marks	Mark band 3: 5–6 marks
A **basic** description of the planning stages to be used in the manufacturing of the prototype.	An **adequate** description of the planning stages to be used in the manufacturing of the prototype.	A **comprehensive** description of the planning stages to be used in the manufacturing of the prototype.
Shows **limited** understanding of safety considerations.	Shows **some** understanding of safety considerations.	Shows a **detailed** understanding of safety considerations.
Completion of the production plan is **dependent** upon assistance or help from other sources.	Completion of the production plan is carried out with **some** assistance or help from other sources.	Completion of the production plan is carried out **independently**.

Top tips

- It is useful to map out an overall picture of the project in block form, identifying the main stages – for example, by using a flowchart.
- Use planning tools to manage your time and resources effectively – for example, a Gantt chart.
- Plan each stage so that you know what you should be doing, where, when, how and with what resources – for example, create a planning table.
- Make sure the materials, components, tools, equipment and processes you plan to use will be available for when you will need them (prepare a parts list and a cutting list).
- Be sure to check all risk assessments and link them with appropriate PPE to your planning.
- Use technology to support your prototype development (CAD and CAM).
- Plan checkpoints and testing, so that you can be sure you are on the right track.
- Evaluate your testing to help you to make sound decisions.

Model assignment

Review hand controllers for a games console. In particular, note the ergonomic form and how they have been designed for hand-held use.

Following your review, you should carry out the following tasks:

- Generate a 2D and a 3D sketch for a new creative design for a Bluetooth games console hand controller.
- Label the sketches to identify parts and dimensions, and annotate the sketches to explain how the shape and form of the hand controller could be produced using craft-based modelling techniques.
- Draw up a cutting list of the materials required to make a concept prototype model.
- Produce a detailed planning table for modelling the concept prototype model using craft-based modelling techniques, to include the following headings:
  - Step
  - Description of the stage
  - Materials
  - Tools, equipment and processes
  - Health and safety/PPE
  - Quality control management
  - Time (min)
  - Additional information
- Generate a 3D virtual model of your design and save it with a manufacturing file extension (such as .stl, .fnc, .dwg, .dxf).

Example candidate responses

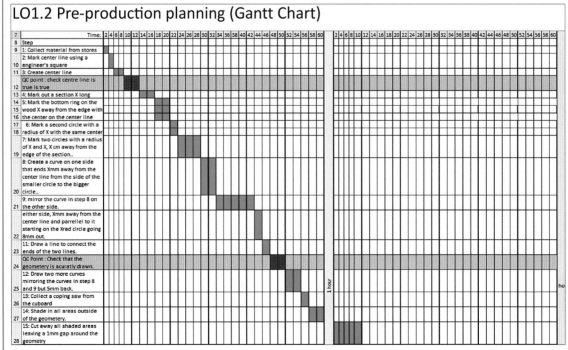

8	1	Component 8	Injection moulding	ABS
7	1	Component 7	Injection moulding	Acrylic
6	1	Component 6	Die Cast	Silicone
5	1	Component 5	Die Cast	Silicone
4	1	Component 4	Injection moulding	ABS
3	1	Component 3	Injection moulding	ABS
2	1	Component 2	Injection moulding	ABS
1	1	Component 1	Injection moulding	ABS
Item	Qty	Part Number	Description	Material
Parts List				

Exploded drawing of the final design:

On the left, I have included an exploded working drawing of my bike light so you can quite clearly distinguish each component and how they are all assembled and connected to one another. In the bottom right hand corner, I have also included a numbered list of each component and the material and process that makes each component.

This exploded view helps a person to clearly visualise the overall finished product of the bike light. It is also incredibly helpful in the fact that you can clearly see the shape of the separate components that need to be produced, and how these components are connected to other pieces.

These pieces of information are a necessity when designing a product.

Use of planning tools – parts list: This example shows a high-quality exploded drawing produced by CAD as well as a parts list. This is useful information that is required for planning resources.

A cutting list for materials would also be useful here.

LO1.2 Pre-production planning (Gantt Chart)

Use of planning tools – Gantt charts: This is a good example of a Gantt chart. It provides detail of processes and time management, and it recognises the need for checkpoints (red lines).

The candidate hasn't added the actual length of the time periods, for example 15–30 minute slots.

LO1.2 Prototype Pre-Production Plan

Step	Process	Tool and equipment needed	Material	Health and safety	Quality control	Additional Information
120	Now mark two lines that are perpendicular from the centre line from the end of the lines in step 110 to the sides of the circle.	Pencil	""	-	Use a steel rule to check all measurements.	
130	Clamp the block in a vice and cut along the outside geometry leaving a 1-2mm gap.	Coping saw	""	Be careful not to clamp your hand in the vice or cut yourself with the saw.	See if the cut is on the outside of the lines.	
140	Use a file to remove almost all of the excess material and then use 240 grit sand paper to finish up the outside finish.	File	""	Be careful not to file your self as it can cause irritation and lacerations.	Use a rule to check the dimensions of the con-rod are still in tolerance.	
150	Get a 24mm and 16mm drill bit, forester or hole saw are good, and drill the corresponding hole.	Pillar drill	""	Make sure you tie back long hair so it doesn't get caught in the chuck. Wear goggles to stop dust getting into your eyes.	Make sure that the holes are centred and there's a solid piece of wood under the piece to stop delamination.	
160	File away 5mm on either side of the square part being careful not to touch the circle.	File	""	Be careful not to file your self as it can cause irritation and lacerations.	Do this carefully to prevent delamination and measure after.	
170	Then coat the exposed edges of the con-rod in a thin layer of PVA to stop de-lamination.	PVA glue	""	-	Be careful not to touch either top or bottom faces and to cover all the edges.	
180	Mark two points both 4mm away from the edge on the rectangle extrusions centred in the other orientation.	Pencil	""	-	Make sure the mark is in the correct place using a rule.	
190	Drill a 3mm hole through each point straight down.	Pillar drill	""	Make sure you tie back long hair so it doesn't get caught in the chuck. Wear goggles to stop dust getting into your eyes.	Make sure the hole is straight and that you drill it slowly so it doesn't delaminate the piece.	
200	Now use a coping saw to cut the 25mm circle in two pieces creating a split con-rod. Your MDF con-rod is now complete.	Coping saw	""	Be careful while cutting so you don't slip and cut yourself.	Check the cut is straight and through the middle of the circle.	

Use of planning tools – planning table: This example shows one page of two detailed and systematic planning tables that list at least 20 steps.

There are no test and evaluation points included in the planning table.

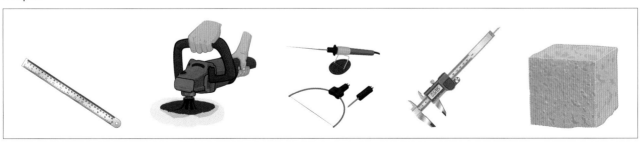

Resources when making a prototype – tools and equipment: This example shows detailed knowledge and understanding of the tools required – what they are called, what they look like, what they are used for and any potential health and safety issues they may present.

Prototype production

Getting started

Search YouTube for 'Prototyping and Model Making – Students of Product Design Episode 5' and watch the video.

List ways that a prototype model can help in the development of a high-quality product.

Following a review of the engineering drawings, the model-maker will consider the modelling methods that could be used to construct the prototype model. Models can be developed from:

- **sheet** materials, such as card, thermoforming polymers or foamboard
- **block** materials, such as expanded polystyrene, wood or metal.

Key terms

Sheet A flat material up to approximately 10 mm thickness.

Block A rigid piece of material that is supplied with relatively flat surfaces.

Laminate Layers of material that have been compressed and bonded together by an adhesive.

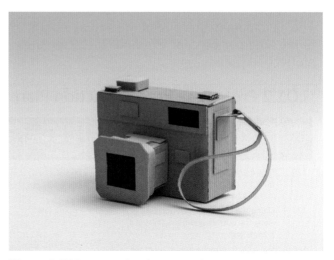

Figure 3.45 An example of card modelling

Card is a sheet material that is used widely in model-making, from early sketch modelling to low-cost undeveloped functioning models. Card is inexpensive and available in a range of thicknesses, colours and types, including single layer, multilayer, corrugated, laminated and cartonboards. Card can be cut, shaped, folded and joined easily using adhesives or mechanical fixings, and finished to a reasonably high quality.

There are several types of polymer available in sheet or block form that lend themselves well to prototype modelling. They include thermoforming polymers, expanded foam polymers and polymers with a low melting point. Sheet polymers can be formed to create hollow structures that can contain components, whereas expanded foam polymers tend to be solid structures with no internal components. Polymers can be used in both functional and concept prototype modelling.

Foamboard is a fairly rigid, lightweight sheet material of **laminate** construction; a layer of expanded polystyrene is sandwiched between two layers of card. Foamboard is available in a range of thicknesses and can be cut and joined similarly to card.

Wood is a very versatile natural material that is available in a range of thicknesses and types, from very thin sheet decorative **veneers** to large solid blocks of softwood (coniferous) or hardwood (non-coniferous). Wood can be shaped, formed, joined and finished to a very high quality; some woods are easier to work than others and so more suitable for prototype modelling.

The type of metal used for prototype modelling will depend on its application – whether it is being used for structural or aesthetic reasons. Sheet and block metals can be shaped and formed, joined and finished, but working metals can be difficult, and robust and dedicated tools and equipment are often required.

For prototype models that require an electronic circuit, **breadboarding** can be a quick method for constructing and testing circuits. There are two types of breadboard: solderless temporary circuit breadboards and solderable permanent circuit breadboards. With solderless breadboards, electronic components and jump wires can be inserted by a push fit into a polymer board that has lines of parallel connector strips under the top polymer layer. This method of circuit construction is often used in early circuit development for testing and trialling ideas and theories; they can be modified very easily and connected to a wide range of input and output devices. The model-maker does need to be very careful not to accidentally dislodge components from breadboard connector strips or the electronic circuit may not function as intended.

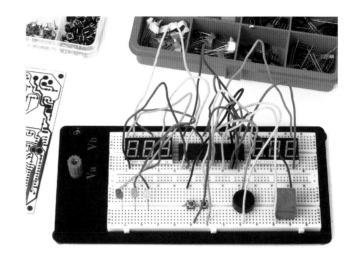

Figure 3.46 Solderless breadboard with electronic components

Solderable breadboards are similar to solderless breadboards in that they also have lines of parallel connector strips. Components and jump wires can be inserted through holes in the breadboard but connections are soldered to form a permanent and more durable circuit board. Solderable breadboards can be cut to an appropriate size and can be connected to a wide range of input and output devices.

Breadboards can be a time- and cost-effective modelling alternative to printed circuit boards (PCB). Final prototype models will often have dedicated PCBs that have been designed and error-proofed on a computer before being manufactured.

Key terms

Veneer A thin sheet of wood that has been sliced off a tree trunk or log. Often used for decorative finishes.

Breadboarding The construction of an electronic circuit on a board (solder or solder-free) using jumper wires to transfer voltage around the breadboard.

Figure 3.47 Solderable breadboard circuit

Occasionally, prototype models have shape or geometry that can be produced virtually on a computer monitor using CAD but that are difficult to achieve for a **physical prototype** using traditional hand tools and machines. In these situations, **additive-manufacturing** processes such as **3D printing** can be an alternative production technique; designers produce designs directly from CAD data.

Traditionally, the process of manufacturing a component has involved subtracting material from a larger piece, leaving behind the shape of the desired component – for example, machining uses cutting tools to remove material from a block. Additive manufacturing works differently; it builds a component by adding material layer by layer.

Additive manufacturing can build components in a wide range of materials, including polymers and metals. Unlike subtractive processes, it is not restricted by where a cutting tool can reach and so can make previously impossible complex shapes.

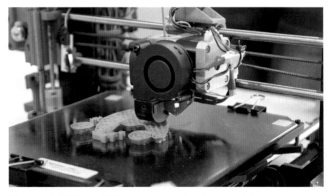

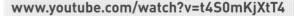
Figure 3.48 An additive-manufacturing machine produces a component by adding material layer by layer

In additive manufacturing, a 3D computer model of the component is created first. This model is processed by software that slices it into layers. Each of these layers is then produced by the additive-manufacturing machine until the 3D component has been created.

Additive-manufacturing technology was traditionally used to produce components that could be used as prototypes. However, as the technology has improved, it has started to be used to produce final-production components. This gives designers far more freedom in the shapes and geometry they can create.

Selection and use of appropriate materials, processes, tools and equipment to produce a prototype

Materials

The properties and characteristics of materials are often overlooked when the engineer plans the process of producing a prototype. Selecting the most appropriate materials for the prototype should not be left to chance – instead, think carefully about how they can be cut, shaped, formed and manipulated to produce an outcome with the dimensions and appearance given in the engineering drawings. Where a particular aesthetic is required, the engineer should select materials that will enable this to be achieved.

The range of materials that can be cut, shaped, assembled and finished is vast and will vary from workshop to workshop, depending on the available facilities. Commonly available materials for prototype modelling include polymers, woods, manufactured boards and metals.

Polymers

Polymers are a group of materials that are made from chains of large molecules. They are commonly named plastics, which are a type of polymer. The types of polymer that lend themselves well to prototype modelling and making are thermoforming polymers, thermosetting polymers and expanded foam polymers.

Thermoforming polymers can be formed and moulded using heat processes, and cut and shaped reasonably easily using a range of hand and machine tools. They are generally available in sheets, rods, extruded sections, tubes and rigid blocks, at relatively low cost. Most thermoforming polymers are available opaque or in a colour, and they often require only limited additional finishing. Processes for shaping and forming these polymers include line bending, injection, extrusion and blow, and rotational moulding, vacuum forming and laser cutting. They can also be placed in an oven and heated to a temperature when they become pliable for hand-moulding processes. Thermoforming polymers can be moulded to form thin-wall structures suitable for casings for electronic circuits, motors and mechanisms.

Prototype modellers tend not to use thermosetting polymers as much because they are more difficult to work. They are available in a range of colours and are generally used in applications with electricity or heat because they are poor conductors of both. Typical uses include electrical switches and plugs, cooking utensils, and oven and pan handles.

Expanded foam polymers, such as polystyrene, are useful lightweight modelling materials. They are available in either sheet or block form and can be cut and shaped using traditional modelling hand tools or a hot wire cutter. They vary in density; the more rigid, high-density versions can be machined by drilling, milling or turning on a centre lathe. Expanded foam polymers are often used for block modelling concept prototypes. They are not easy to join using traditional mechanical fixings but they can be joined by adhesives and adhesive tapes. They are only available in a limited range of colours but can be sealed, rendered and finished to a high standard.

Polymers with a low melting point, such as polycaprolactone (PCL), can be useful for modelling smaller components and products. An example is Polymorph, a non-hazardous and biodegradable polyester. When heated in water at around 60 °C, it melts, and it becomes solid again once cooled. When molten, it can be moulded to shape by hand; as a solid, it can be machined – for example, drilled and threaded.

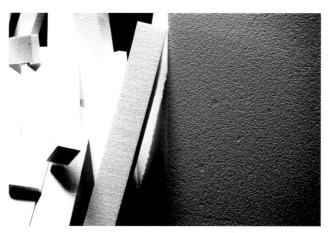

Figure 3.49 Expanded foam polymers

Woods

Wood is a natural resource that lends itself well to prototype modelling and making, but it is not always easy to work with. Each type of wood has natural properties and characteristics that need to be considered before modelling, such as weight and strength, resistance to cutting and abrasion, **grain** structure and texture, and how it is affected by humidity, to name but a few.

Some woods, such as the very lightweight balsa and the soft, even-textured jelutong, make excellent woods for prototype modelling. The advantages are:

- They are straight-grained timbers that cut easily and do not produce particularly sharp splinters.
- They can be machined, so long as cutting tools are sharp.
- Thin veneers can be laser cut.
- They have a fine grain that can be sanded for an excellent smooth surface.
- Once sealed, they finish well with wax, varnish or paint.
- Once finished, they could appear to have the visual characteristics of any number of materials.

Wood veneer is a thin slice of wood that can be bonded to the surface of a manufactured board to improve its appearance. Typical applications include **carcases** and table tops where a manufactured board provides structural strength and veneer the visual impact. Natural woods can be sealed and finished to a very high quality.

Figure 3.50 Balsa timber with its fine grain structure is an excellent modelling material

Manufactured boards

Manufactured boards – including particle boards, laminated boards and fibreboards – are timber-related products that provide an effective alternative to natural wood. If a wood-grain finish is required from a material with good strength characteristics in all directions, then a laminate board, such as plywood, could be used. However, if a smooth, even surface is preferred for a carcase, then a polymer-coated board could be more appropriate, although the ends would require additional attention.

Fibreboard is a good modelling material because it can:

- be cut and shaped using a wide range of hand and machine tools
- be joined using wood adhesives
- take wood and self-tapping screw across the fibres
- be brought to a very smooth finish across all faces, which when sealed can be rendered to a high-quality finish with paint.

Key terms

Grain Natural alignment of fibres seen in a cut surface of wood.

Wood veneer Very thin slice of wood up to approximately 3 mm thick.

Carcase A hollow structural enclosure for containing an object, such as a motor, electronic circuit or mechanism.

Figure 3.51 Fibreboard being clamped and cut

Metals

It can be a challenge to use metals for modelling and making prototypes because they tend to be difficult to manipulate accurately.

Low-carbon steels are **malleable** and, therefore, can be shaped and formed hot or cold. They can also be machined, although cutting tools will quickly heat up and wear out unless an appropriate coolant is used.

Thin-sheet carbon steels bend and fold relatively easily, and they can be joined permanently by riveting or by heat processes such as welding and brazing, or non-permanently using mechanical fixings.

Key terms

Malleability The ability of a material to be shaped or deformed by compressive forces (such as hammering or pressing).

Ferrous material Material that contains iron.

Corrosion Gradual destruction of materials due to a reaction with their environment or chemicals.

Alloy A mixture of two or more metals.

Casting Process where solid material is heated until it turns into a liquid and is then poured into a mould; once cooled, the material will have taken the shape of the mould.

Anodised Coated with a protective oxide layer using an electroplating process.

Carbon steel is a **ferrous material** that will corrode in moist environments; therefore, a finish is often required to enhance its aesthetic appeal. Once sealed, carbon steels can be rendered to a very high-quality finish. Stainless steel is resistant to **corrosion**, can be polished to a high-quality finish and is often used where a metallic finish is required. Most carbon steels are available in a wide range of thicknesses, forms and profiles, such as bar, round and extruded sections, as shown in Figure 3.52.

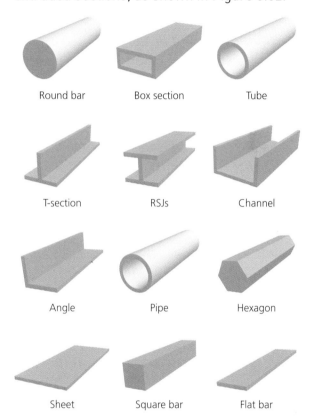

Figure 3.52 Different profiles of carbon steel

Aluminium is often used as an alternative to carbon steel. It has a much lower density than other metals and is often mixed with other metals to form an **alloy** to improve its properties, such as strength. It can be machined, shaped and finished to a very high quality and is an excellent material for low-temperature **casting**. For protection against the elements, it requires no specific coating; however, it can be **anodised** or coated if a rendered finish is required. Aluminium is often used where a long-lasting and non-corroding metallic finish is required. Like carbon steels, it is available in a wide range of thicknesses and forms.

Thin-sheet aluminium can be formed and shaped cold, but thicker sections tend to fracture if folded too much. Aluminium components can be joined permanently by riveting or welding, or non-permanently by mechanical fixings.

Activity

Produce life cycle analysis (LCA) diagrams, from extraction through to end of life, for woods, metals and polymers.

Tools, equipment and processes

Once a decision has been made about the type of prototype to be developed and the materials that could be used to make it, the next stage is to identify tools, equipment and processes that could be used for its manufacture. Table 3.19 shows the key processes and many of the tools and equipment that could be used.

Table 3.19 Tools, equipment and processes that could be used for prototype modelling

Process	Tools/equipment
Marking out	Pencil, marker pen, stencil, flexible curve, steel rule, engineer's square, divider, odd-leg calipers, templates, stencils, laser cutter
Wasting	For concept prototype modelling: a modelling knife, scalpel knife, hand saw, scroll saw, band saw, circular saw, linisher, bobbin sander, router, wood lathe, hand drill, pillar drill, hot-wire cutter, laser cutter
	For functional prototype modelling: a centre lathe, milling machine, grinding machine, CNC machine (such as lathe, milling machine, router, machine centre, laser cutter)
Finishing	Glass paper, wet and dry paper, paint, oil, varnish, wax
Additive manufacturing	Rapid prototyping (3D printing)
Assembly	Adhesive, permanent fixings, non-permanent fixings
Quality control	Micrometer, depth gauge, vernier calipers, vernier height gauge, go/no-go gauges, thread gauges, radius gauges, digital readout display

The models shown in Figures 3.53 and 3.54 are two very different examples of prototype development and manufacture. Both prototypes have been made using a range of materials, tools, equipment and processes (including computer-controlled and rapid-prototyping processes). The prototype hot-melt glue gun has been assembled to form a concept prototype model; it has no internal components and cannot be connected to the mains electricity supply. Following evaluation, the prototype can be finished using appropriate finishing processes.

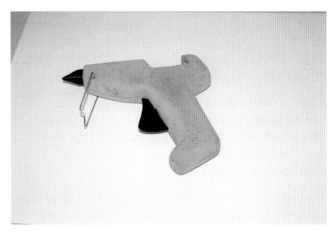

Figure 3.53 A prototype hot-melt glue gun

Looking at Figure 3.53:

- The prototype has been made from two pieces of fibreboard that have been cut out and joined using an adhesive. They have then been shaped using craft-based cutting and shaping techniques.
- Holes have been drilled using a hand drill and a pillar drill.
- The fibreboard has then been sanded and finished for sealing.
- The wire stand has been shaped and formed using a cold-forming technique.
- The trigger has been cut from acrylic sheet using a laser cutter.
- The nozzle has been produced by 3D printing (rapid prototyping).

Figure 3.54 Prototype development using a range of materials and processes

The prototypes shown in Figure 3.54 demonstrate the development of a piston connecting rod design. A range of different materials and processes have been used to develop a solution, including rapid-prototyping 3D printing. The final prototype was produced from only one process and one material, but it was the conclusion of trialling and testing with various materials, tools, equipment and processes.

- A basic block model prototype has been shaped from fibreboard. Modelling could also have included manual milling to remove excess material, or CNC routing.
- A laser-cut model demonstrates the use of computer-controlled processes. This model was built up in layers to achieve the required thickness.
- The third prototype model was produced by rapid prototyping. The 3D-printed model displays detail and characteristics that meet specification criteria.
- The final prototype's shape, size and proportion have been developed through progressive stages of modelling and testing.

Table 3.20 identifies some tools, equipment and processes that can be used to cut and shape materials.

Table 3.20 Tools and processes used to cut and shape materials

Hand saw	Many different types of hand saw are available depending on the application, including hack saw, coping saw, crosscut saw, panel saw, rip saw and tenon saw. All have metal blades and angled teeth for cutting and removing waste. Most require only one hand around the handle and cut on the forward stroke of the blade through the material. However, a coping saw cuts on the pulling stroke. The saw should be held lightly and comfortably in the hand and controlled as it cuts through the material.
Jigsaw	This is a portable powered saw that can be used to cut shapes from sheet material by moving back and forth/in a straight line. It has removable blades that can be changed to suit the type, thickness and desired finish of the material being cut. A jigsaw cuts on the down stroke, and space for the blade is required below the material being cut. Jigsaws are not very accurate cutting tools because the blade is only secured at one end.
Scroll saw	This is a fixed-position machine saw that can be used to cut small shapes from hand-held sheet material. A range of blades are available for cutting different types of material. Similar to a jigsaw, the blade cuts back and forth, but the scroll saw cuts on the up stroke, so it is important to have the material firmly located under the hold-down.
Pillar drill and hand drill	Fixed-position pillar drills and portable hand drills cut a relief below the surface of a material, such as a countersink or counterbore, or a hole using a drill bit. Pillar drills are designed to drill along a vertical axis. Many pillar drills have a table that can be rotated to allow work to be mounted at an angle for drilling. Portable hand drills provide an opportunity to drill into materials that may not be able to be drilled easily on a pillar drill – for example, due to location, size or shape.
Hot wire cutter	A hot wire cutter is a useful tool for cutting expanded polystyrene. It can be hand-held or table-based and consists of a thin, taut metal wire which is heated by an electric current to approximately 200 °C. As polystyrene is lightly pressed against the hot wire it melts, forming a cut line just in advance of contact.

Laser cutter	A laser cutter is a computer-controlled machine that can be programmed to etch or cut through sheet materials, such as paper, card, polymers, woods, boards and some thin sheet metals.
	Sheet material is placed onto a table in the machine and a powerful laser beam vaporises a path through the material as it passes over it. The laser beam follows a path determined by a software driver program, according to a drawing produced using CAD software.
CNC router	A CNC router is a machine that can be programmed to cut and shape a wide variety of rigid materials along both horizontal and vertical coordinates. A mill-type cutting tool is positioned in a chuck and it rotates at high speed. The cutting tool follows a path determined by a software driver.

Forming and bending

The shape of sheet materials can be modified by bending and stretching rather than by removal of material. The material is forced to change shape by the application of pressure, as represented by the arrows in Figure 3.55. It is possible to do this to woods, composites, metals and polymers. However, some materials are more easily formed than others. For example, wood can be formed, but moisture in the form of steam or water is often required to soften the grain to avoid it breaking.

Figure 3.56 A coffee table produced by veneer laminate moulding

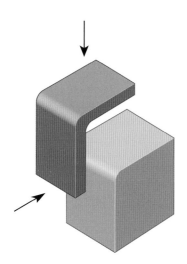

Figure 3.55 A simple 90° former

An alternative to bending wood is to produce a laminate of thin veneer layers of wood that are bonded together by adhesive as they are compressed to the required shape on a **former**. Once the adhesive has dried, the formed laminate will be permanently shaped, as can be seen in the coffee table in Figure 3.56. This type of forming requires a former above and below the material being formed, so that accurate shaping can be achieved.

Key term

Former Device over or around which materials can be formed.

Most thin-section malleable metals – for example, aluminium, copper, brass, gold, silver and some carbon steels – can be formed and shaped without the requirement of heat, although accurate forming may require a metal folder machine. Thicker metal bars and rods cannot always be formed while the metal is cold, so processes may be required to heat the metal to a temperature at which it becomes sufficiently workable.

Many thermoplastic polymers can be formed and shaped relatively easily and quickly by the processes shown in Table 3.21.

Table 3.21 Methods of forming and shaping thermoplastic polymers

Method	Description
Line bending Strip heater	Line bending is the process of heating a polymer such as acrylic or **high-impact polystyrene (HIPS)** along a line to produce a simple fold. The polymer will be marked on both sides, along the line of the required fold, and then suspended above the strip-heater element along that line; heating both sides of the polymer equally helps to achieve an even fold. Once the polymer is at the required temperature, it can be transferred to a former or jig for accurate folding.
Press moulding Yoke Guide pegs Plug Finished moulding	Where more complex shapes are required, such as a casing or an end cap, the **yoke and plug** press-moulding process can be used to form and shape thin-sheet thermoplastic polymers such as acrylic and HIPS and foamed polymers such as Plastazote©. The sheet polymer is placed in an oven and heated until it becomes fully pliable. It is then placed between the yoke and the plug and compressed. Once the polymer has cooled, it can be removed from the press mould and excess material can be removed.
Drape forming Heated acrylic Plastic pipe Cloth Block for vice	Drape forming is used to form large, curved shapes from a thermoplastic polymer or a foamed polymer. As with press moulding, the sheet polymer is heated in an oven until pliable, then draped over a former and compressed by a piece of cloth.
Vacuum forming Clamp · Heater Plastic sheet · Mould Platen Heat · Raise platen · Vacuum	Vacuum forming is used to shape and form thin-sheet thermoplastic polymers. A mould is located within a vacuum chamber and a polymer, such as HIPS, is clamped to a frame above the mould. The polymer is then heated from above; once it is uniformly pliable, it is lowered onto the mould. A vacuum pump is turned on to remove air from between the polymer and the mould. The polymer is tightly drawn down over the mould and then left to cool. Once cooled, the polymer will have taken the form of the mould.

Method	Description
Rapid prototyping (3D printing) 	When producing a prototype by craft-based modelling processes, it is not always possible to create a component or product as accurately or as detailed as shown in the engineering drawings. Rapid prototyping is an additive-manufacturing process where a CAD-generated 3D engineering drawing can be interpreted and turned into a 3D physical prototype. 3D printing is a manufacturing process that deposits hot filament polymer, such as **acrylonitrile butadiene styrene (ABS)** or the natural **polylactic acid (PLA)** polymer, layer by layer to form a component or product. Products may be formed as solid constructions or can include multiple components. Some of the more sophisticated machines have multicoloured polymers that can be deposited one after another, so that each component within the product can be a different colour.

Key terms

High-impact polystyrene (HIPS) Rigid thermoforming polymer with high-impact strength.

Yoke and plug Two parts of a mould that can be used to shape pliable materials.

Acrylonitrile butadiene styrene (ABS) Tough, rigid thermoforming polymer with high-impact strength.

Polylactic acid (PLA) Natural polymer made from corn starch or sugar cane.

Template Shaped piece of (usually) paper or card that can be used to test a profile or to mark around.

Pattern Reusable, shaped piece of robust material that can be used to guide a manual machine-tool operation.

Mould Device for producing a 3D form in the shape of the desired outcome.

Activity

Produce a list of the tools, equipment and processes that have been identified in this unit.

To the side of each tool, equipment and process mark with a ✔, a ? or a X:

✔ indicates the ones you know and are confident of working with.

? indicates the ones you are unsure of, or not too confident of working with.

X indicates the ones you do not know.

Assembly methods

When preparing and assembling materials, engineers often need devices such as **templates**, **patterns**, jigs, fixtures, formers and **moulds**. These devices enable accurate marking, machining and assembly. For one-off or very low-volume production, they will often be basic, low-cost devices, whereas for large batches and high-volume production, dedicated robust and expensive devices will be manufactured.

Templates

A simple template can be a very useful aid, particularly when marking out or locating position for the assembly of irregular shapes or profiles. A template for one-off use is generally made from paper or card and stuck onto the surface of the material or component to be machined or assembled. The template may have the position of holes, slots, profiles or other detail marked onto it or cut out. These can be markings to aid machining or assembly that may be difficult to accurately produce directly on the material or component by standard marking-out methods. Templates can be produced from hand drawings or from CAD/CAM processes, such as laser cutting.

Figure 3.57 shows a card template with hole markings. Due to there being limited square faces to mark from, a template would be a useful aid to mark the position of the holes for drilling.

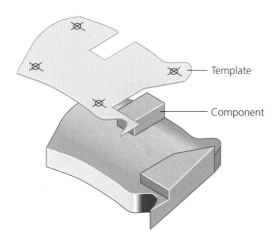

Figure 3.57 A simple template with hole markings

Patterns

A pattern is similar to a template but made from a more robust material, which can be used multiple times. Patterns can be used for marking out because they can be drawn around, but more often they are used for machining, where a cutting tool will follow the profile of the pattern. A good example of profile cutting using a pattern would be to shape an edge using a **router** and pattern. The pattern is clamped to the workpiece and the router is guided along the profile to produce an exact replica of the pattern's profile, as shown in Figure 3.58.

Figure 3.58 Profile cutting using a pattern and router

Key term

Router High-speed hand-held or table-mounted rotary cutter.

Jigs

Jigs can help the engineer when performing a machining operation or when simply assembling parts. The purpose of a jig is to enable accurate location – for example, it could be used to precisely align components so that they could be drilled and fixed together. For a one-off task, a jig can be functional but undeveloped; however, for repeated tasks, it is sensible to manufacture a dedicated jig to enable accurate machining every time.

Figure 3.59 shows a dedicated machining jig. Each time a component is mounted onto the jig, it will be in the exact position for accurate machining.

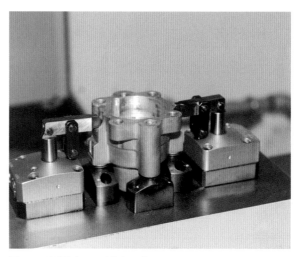

Figure 3.59 A machining jig

Fixtures

A fixture can be used when multiple identical components are to be machined one after another using the same methods. It is a device that holds a workpiece securely in a specific location for machining operations.

Moulds

A mould is a device for producing a 3D form in the shape of the desired outcome. It is possible to shape:

- around the outer surface of a mould – for example, during vacuum forming
- through a mould – for example, during extrusion moulding
- from the inside of a mould – for example, during rotational moulding, blow moulding and compression moulding.

Where the moulded product is a shell structure, a mould will be made from two or more parts, as with plug and yoke moulding, injection moulding and die casting.

Joining and fixing

The joining and fixing of components is one of the final stages within the modelling process, and the techniques used often depend upon the type of prototype and whether the joining/fixing method is to be seen on the final product:

- On a model where the fixing will be visible as part of the final product, such as a bicycle light clamping bolt, the same method could be selected for both a concept prototype and a functional prototype.
- Where the fixing method is not meant to be seen, such as moulded snap-together location clips on a television remote-control casing (Figure 3.61), an alternative fixing method such as an adhesive can be used because the snap-together clips will not be seen, only a closure line around the casing.

When an adhesive is to be used, the engineer must check its compatibility with the materials being glued together and follow the directions for use on the safety data sheet.

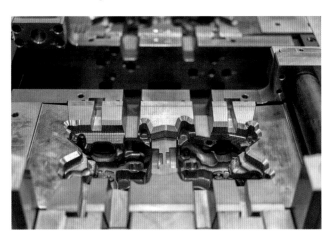

Figure 3.60 A high-precision die mould

Figure 3.61 Moulded snap-together location clips

Table 3.22 Types and uses of various adhesives

Adhesive	Type	Materials it will join
Glue stick	Natural ingredients, including water, sugar and potato starch	Paper, thin card
PVA	Water-based synthetic adhesive	Thicker card, corrugated cardboard, foamboard, wood
Epoxy resin	Two-part adhesive; resin and hardener mixed in small quantities	Most modelling materials
Synthetic resin	Urea/formaldehyde powder and water mix	Woods

Adhesive	Type	Materials it will join
Superglue	Fast-curing single-component liquid	Most modelling materials on contact
Contact adhesive	Rubber-based solvent adhesive; low-tack also available for non-permanent joints	Most modelling materials (often used on sheet materials but not for use on polystyrenes)
Hot-/cool-melt glue stick	Quick-setting, thermoplastic adhesive, becomes liquid when heated	Most modelling materials
Grab adhesive (such as 'No nails')	High-strength, water-based adhesive	Most modelling materials

When joining the same type of polymers, exclusive adhesives are available – for example, polystyrene cement for polystyrene and dichloromethane methyl methacrylate (Tensol cement) for acrylic.

Where mechanical fixings are to be used, there are many different permanent and temporary types to choose from. For each application, the most appropriate fixing should be selected considering factors such as material, size, strength and ease of use. Where a fixing will be noticeable, visual characteristics such as colour and finish may also be considered; where the fixing should not be seen, methods of concealing the fixing may be required.

Table 3.23 Non-permanent fixings

Type	Description
Nails	• Quick method of joining to wood, normally by hammering • Available in a range of materials, sizes and types, such as round wire nails, oval wire nails, masonry nails and panel pins
Wood screws	• Strong tapered steel or brass fixings, with a shaped head for joining to wood • Available as cross head or slot across a wide range of sizes
Self-tapping screws	• Reasonably strong fixings when used with polymers, where the steel screw is turned into a hole and cuts its own thread as it goes in • When used for joining sheet materials, they can produce a very strong joint if used with captive nut fittings • Available as cross head or slot in a wide range of sizes
Machine screws	• Strong threaded fixings, tightened using a screwdriver • Used for joining materials in conjunction with a nut or where one piece has an internal thread • Available as cross head or slot in many different thread sizes, lengths and materials
Socket-head screws	• Strong threaded fixings, tightened using an **Allen key** • Used to join materials similar to machine screws • Available in many different thread sizes, lengths and materials
Hexagon bolts	• Strong threaded fixings, tightened using a spanner • Available in many different thread sizes, lengths and materials

Nuts and washers	• Internally threaded fittings that can be used in conjunction with machine screws, socket screws and hexagon bolts to form a tightly compressed joint
	• As the nut is turned up the thread, it compresses the materials being joined against the head of the screw or bolt
	• Nuts are available in different thread sizes and types, including hexagonal nuts, square nuts, wing nuts, nylon locking nuts and castle nuts with a split pin
	• Washers are generally used ahead of a nut, and there are four basic types:
	  • flat washers, designed to spread the load of the nut across a larger cross-sectional area
	  • shake-proof, tab and spring washers, designed to prevent a nut from vibrating and turning loose.

Knock-down (KD) fittings	• Strong fixings, such as cam fittings, connector blocks and cross dowels • Often used to join wood or board sheet material face to end, such as flat-pack furniture joints

Table 3.24 Permanent fixings

Type	Description
Soldering	• Soft-soldering: relatively low-temperature (200–300 °C) method of joining metals using a fusible metal alloy (solder); for example, used to join electrical components to a printed circuit board • Hard soldering and brazing: method of joining similar or dissimilar metals using heat (typically from a gas torch), flux and a compatible filler material; soldering processes happen at temperatures below 450 °C, while brazing happens at temperatures above 450 °C; for example, used in jewellery making
Welding	• Electrical arc welding (MIG [metal inert gas], TIG [tungsten inert gas]): process for joining metals at high temperature using an electrical spark; a filler is often used to add material to the joint • Gas/oxyacetylene welding: process for joining metals, using a mixture of acetylene and oxygen to produce a high-temperature flame • Other forms include electrical spot welding for joining sheet metals, such as car bodywork panels, and polymer hot-air welding, melting polymers to form a welded joint
Bonding	• Process of fixing using a high-strength compound, similar to two-part adhesive fixing • Bonding occurs as a result of solvent evaporation or through curing via heat, time or pressure • Compounds are available for joining a range of materials, one part of the compound being a hardener
Riveting	• Pop-rivet: fast and efficient blind fastener method that only requires access to one side of the joint; generally used to join to a sheet metal • Simple headed pin rivet: generally used to join two or more sheet metals; unlike pop-riveting, access is required to both sides of the rivet so that it can be compressed
Folding	• Method of joining sheet metals by folding them together (the tighter the fold, the stronger the joint)

Key term

Allen key Hexagonal-shaped tool that fits in a socket-head screw.

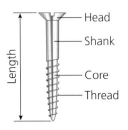

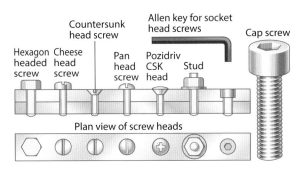

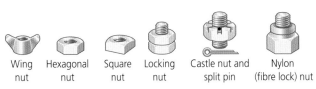

Figure 3.62 Examples of mechanical fixings

Activity

Research polymer moulding.

Sketch and label the processes for heating thermoplastics as used when injection moulding, extrusion moulding, blow moulding and rotational moulding.

Record the key stages of modelling the prototype

The completed prototype model is visible evidence of the knowledge and understanding that have been gained from selecting appropriate modelling methods, materials, tools, equipment and processes, and evidence of skills gained and used. You need to provide evidence to show how the prototype was made in the form of an annotated photographic record that shows the key decisions and actions taken at key stages throughout the manufacturing journey, as well as close-up images of the completed prototype model. For example, a record should show how materials were marked out, cut, shaped, formed, assembled and finished. The record should list the PPE that was used, and how tools, equipment and processes were used safely and effectively to produce the prototype model. Your annotations should explain how the prototype was produced independently, or where help and support were required.

It needs to be clear to an assessor how you developed your prototype, from a written plan through to a 3D outcome. Recording this should be systematic and as illustrative as possible, so that notes and images can feed directly into a final presentation of evidence without requiring too much extra information. You may only need a portfolio of 10–12 A4 equivalent pages to demonstrate you have met the criteria.

To record evidence of how the prototype was produced, you could use:

- written notes in a notebook or **log book**
- regular updates to the production plan
- images, photographs or videos.

Key term

Log book Ongoing record of questions, decisions and solutions that can be used as evidence of systematic activity and review.

A notebook or log book is a valuable instrument to record your ongoing thoughts, questions and decisions, times and dates, problems and successes, technical difficulties and solutions, with no detail being too large or too small. You can write notes before and after activities, to record issues and experiences that could otherwise be forgotten over time.

The production plan is a working document that should be accessible and that you regularly review and update. It is the guide to what was planned to be achieved. You should over-write amendments to indicate any changes made to the original plan due to new information, skills, knowledge or understanding. An annotated and amended plan tells much more of a story than a beautiful and clean plan.

Photographs and videos can be used to communicate something that may require many words to explain. They can also be an extremely useful *aide-memoire* (helping you to remember) for when you compile the final record (photographs taken at key stages are particularly useful). You must provide photographic evidence to show that you have produced a prototype.

A very useful method is an ongoing photographic production diary with notes. Using photographs, you should record each key stage with annotations to explain the important details and issues as they develop; sketches can also be added. In the production diary, you should record progress against the production plan, noting changes and updates, and how potential health and safety issues have been appropriately addressed, referencing risk assessments, safe working practices and PPE. The production diary may also link evidence across learning objectives, which will later be developed and expanded to meet assessment criteria. Video evidence can also be a useful diary tool, particularly where it includes spoken explanations of what is being recorded and why.

During the production of the 3D virtual model, the writing of your production plans and the making of the prototype model, you will be observed by a witness, a responsible adult. They are required by the examination board to write an observation record for the assessor that indicates your level of independence or the support you received when you were working on those tasks.

The observer will also comment on your use of appropriate PPE and how safely and effectively you used tools and processes to produce and assemble your outcome. The observer cannot gain marks for you; their observation record can only support what you show in your diary of activities.

You are strongly recommended to compile a detailed record at each stage of the task. Record the key stages of each activity by photographs or video and write short annotations to explain the activity and why it was required, state when you did it, the safety measures you took and how independent you were. You should comment on the quality of your outcome – does it function as intended? You should also comment on the PPE you required and how effective it was.

Activity

Produce a short annotated photographic (or video) diary entry to communicate how to set up, use and then clean down a power tool from the workshop (for example, a hand drill or pillar drill).

Test your knowledge

1 Why is card a suitable material for sketch and early modelling processes?
2 Which sheet materials could be used for making prototypes?
3 How can computer-controlled processes be used alongside traditional craft-based modelling – for example, to make stencils, templates or components?
4 What is the difference between a template and a pattern?
5 Which adhesive could be used for joining acrylic to aluminium?

Assignment practice

Marking criteria

Mark band 1: 1–6 marks	Mark band 2: 7–12 marks	Mark band 3: 13–18 marks
Dependent upon assistance to produce a prototype from a production plan.	Requires **some** assistance to produce a prototype from a production plan.	**Independently** produces a prototype from a production plan.
Dependent upon prompts to use PPE equipment when working with tools, machines, materials, chemicals, finishes and solvents.	Requires **some** prompting to use appropriate PPE when working with tools, machines, materials, chemicals, finishes and solvents.	**Independently** uses appropriate PPE when working with tools, machines, materials, chemicals, finishes and solvents.
Uses tools and processes with **limited** effectiveness to produce and assemble an outcome that partly meets the production plan. The prototype will be incomplete.	Uses tools and processes with **some** effectiveness to produce and assemble an outcome that mostly meets the production plan. The prototype will be mostly complete.	Uses tools and processes **effectively** to produce and assemble an outcome that is of a high quality, accurate and fully meets the production plan. The prototype will be fully complete.
Produces a **limited** record of the key stages of making the prototype.	Produces an **adequate** record of most of the key stages of making the prototype.	Produces a **detailed** and accurate record of the key stages of making the prototype.

Top tips

- Develop a systematic approach to modelling/making, following your plan for making at each stage.
- Use a digital camera or video recorder to record evidence at key stages of you using tools and processes safely and effectively, including computer-controlled processes.
- Provide evidence to show you have:
  - reviewed and selected the most appropriate materials, fixings and finishes to produce the prototype model
  - used tools and processes effectively to produce and assemble an outcome that fully meets the production plan

- provided clear annotated photographic or video evidence of you working safely and effectively, and using appropriate PPE
- used appropriate methods to record in detail all the key stages of making the prototype.
- Provide clear annotated photographic or video evidence of you working safely and effectively and using appropriate PPE.
- Your evidence can be supported by a Teacher Observation Record.

Model assignment

The drawing below shows the prototype for a gate hold-back catch.

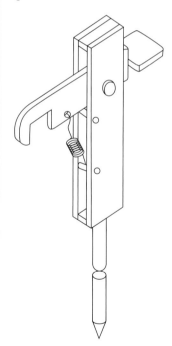

- Select and justify appropriate materials and components that could be used to produce a prototype model of the gate hold-back catch.
- Produce a draft production plan that identifies steps, describes the key stages, identifies appropriate materials, tools, equipment and processes, and considers health and safety, time and quality control devices/processes.
- Using a risk assessment template, complete a risk assessment for the task identifying hazards, risks and control measures (including appropriate PPE).
- Use tools and processes effectively to produce and assemble an accurate and complete prototype model of the gate hold-back catch.
- Use appropriate methods to produce a detailed and accurate record of the key stages of making the prototype model.

Example candidate responses

Identify suitable materials from the list below that could be used as prototyping materials:
pine, clay, acrylic, MDF, PLA, flat section mild steel bar, PLA, plasticine, blue foam and model block.

Material	Description & Properties	Justification
MDF	An easily machinable, inexpensive material that also has a good amount of structural integrity. It's also a safe material as it doesn't produce splinters like other woods and metals.	Since MDF is an inexpensive material that can be machined, it works well as an aesthetic and usable prototype.
Blue Foam	Easily shaped and inexpensive material. It also doesn't have much structural integrity and is very brittle and weak.	Blue foam is even less expensive than MDF and can be shaped with hot wire and craft knives making it great to be used as a rough prototype but as it's a very weak material it can't be used to test the pump.
Flat Section Mild Steel Bar	Tough, hard and strong. However this makes it a harder material to machine than the other materials and not as safe.	Mild steel bar is a good material for a final prototype due to it being close to the material that would be used in the final product as it is tough and can have different finishes applied to it.
PLA	Brittle, has a low melting point and is semi-durable material. It is often used in 3D printing.	PLA plastic is a very good material to use for rough prototypes and physical representations as its low melting point allows it to be 3D printed which creates the prototype over night saving time.

I have chosen MDF to manufacture the main section of my prototype con-rod because it's easy to machine as well as being strong enough to be able to test the design.

In the above example, the candidate has reviewed the available materials and then given a brief justification for one of the materials selected for modelling.

Tools, equipment and machines

Machinery, Tools, Equipment, Materials	Purpose	Health and Safety
Ruler	To measure accurately when marking out a material for cutting and shaping. The reason for using a steel ruler instead of a plastic one, is that the measurements start from 0 (instead of 4mm of normal rulers) and it has increments of 1mm which is great for more detailed measurements.	N/A
Styrofoam	Styrofoam is a lightweight thermoplastic that is easily workable. It can also be recycled making it an environmentally friendly, sustainable material. It will be used to make the prototype considering it is easy to shape using subtractive manufacturing techniques.	When cutting or sanding Styrofoam, consider wearing a mask if you have allergies. This is because Styrofoam produces a lot of debris that could potentially be inhaled or cause allergies to flare up.
Pillar Drill	It is used to drill holes at a high speed whilst keeping the material in place. I will use the pillar drill along with a forstner bit to hollow out the bracket, drilling a hole all the way through the bracket of the prototype.	Wear goggles to protect your eyes. Locate stop buttons on machinery for immediate access if necessary. Be careful with sharp points or blades that could potentially harm people. Use the appropriate speed on the pillar drill to prevent irreversible damage.
Glass paper/Sand Paper	Used to remove material from surfaces, to make them smoother and provide a high quality finish to the product after removing unnecessary layers. It will be used at the end to ensure that the prototype has smooth surfaces and edges and make the model aesthetically pleasing.	Avoid rough, repetitive contact with the skin as the vigorous movements could harm people or cause pain and irritation. Sand paper creates lots of debris – perhaps consider wearing a mask if you have allergies.
Disc Sander	The disc sander provides a quality finish and smooth surfaces and edges to the prototype. It also grinds the material to achieve desired dimensions – trimming the Styrofoam as well as the acrylic screen.	Always wear safety goggles while using the disc sander and turn on the dust collector. Roll sleeves up past the elbow and ensure that no loose clothing will get caught in the disc while operating the disc sander. Locate stop buttons on machinery for immediate access if necessary.

Here the candidate has provided good evidence to show that they clearly understand the assessment criteria. There has been a review to identify the tools, materials, components and adhesive that would be appropriate for modelling the prototype. The candidate has explained the purpose of each tool and material and any potential health and safety issues.

On these slides, the candidate shows a systematic approach to their prototype modelling and making. They have provided clear photographs that show key stages in the process, which have been annotated with detailed explanations. Notice the candidate has demonstrated the link between craft-based materials and computer-controlled processes – a laser cut template.

On this picture you can see evidence of me marking out. I am connecting together both of my conrod with a straight line by using a sharp pencil and steel rule. In my marking out I also used a engineers square and a compass.

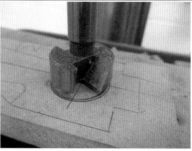

Here are the buttons for the drilling machine you must unlock the red button to be able to start the machine. You then press the green button to start the machine. The switch allows you to change the direction of the drill.

Here I am using a fortsner bit to drill out my centre hole. I wore goggles and an overall to protect me from any hazards, such as dust and wooden splinters from my clothes and eyes. By wearing this equipment I correctly followed the PPE. To secure the MDF I used a metal clamp providing a more fixed hold on material.

Here I am using a coping saw to cut out my design. When I cut my design out I left a 2cm width gap to avoid sanding final product to much making it too small. I wore an overall while doing this to protect my clothes from saw dust produced by sawing.

Research

For further information on materials, processes and engineering-related topics, check out the Technology Student website: **www.technologystudent.com**

Evaluation of the prototype

From a page of initial design ideas, select several ideas for comparison against each other and against the design specification criteria.

- Identify the features and characteristics that make each of the designs a potential solution.

- Produce a **Pugh Chart** to communicate the merits of each design when compared to specification criteria.
- Suggest which design could represent the best potential solution and why.

Evaluation of processes and systems

A high-quality outcome is often the result of first-class evaluation of the processes and systems used to plan and produce outcomes. Businesses seek to review every process and system to see if every aspect of design, development and manufacture can be carried out more efficiently or to a higher quality. An evaluation of planning for making reviews the processes at each stage to see if they were accurate and if they could be improved. The ultimate goal is to deliver a plan that is effective and error free so that, if used again, the quality of outcome may be improved.

When evaluating the quality of processes and systems it is important that, as much as possible, evaluation judgements are **objective**, factual and criteria-based, as opposed to **subjective**, based on personal opinion and possibly biased views. If another person reviews the same outcome and its associated planning documents (charts, planning for making, log books), they should be able to come to similar conclusions.

A **summative evaluation** of planning will objectively look back at the flowcharts, Gantt charts and planning for making documents to identify where there may have been miscalculations or errors that could be rectified

were the whole process to be started again. Where a flowchart is found to have errors, they can be overwritten, or cut and pasted with the corrected flow paths inserted. When reviewing a Gantt chart, it can be marked up with actual timings alongside the planned timings, to help evaluate how realistic the original timings were (see Figure 3.63). Planning for making documents could have additional lines and cells inserted for supplementary information, the aim being to provide a realistic and accurate document.

Key terms

Pugh Chart A design tool for comparing design ideas against specification criteria.

Objective evaluation Appraisal that is based on fact, is reliable and could be repeated if performed by another person.

Subjective evaluation Appraisal based on personal views, which may include bias.

Summative evaluation Appraisal at the end of a series of processes.

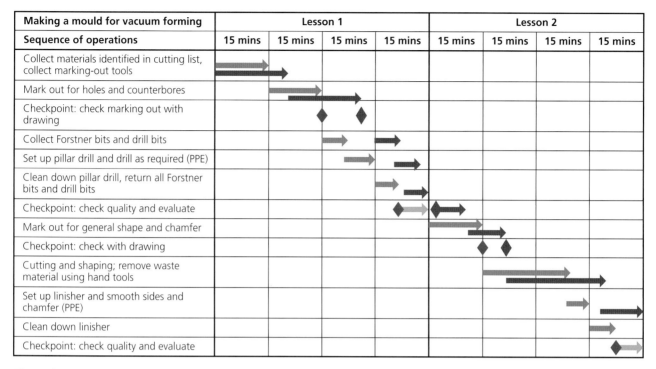

Making a mould for vacuum forming	Lesson 1				Lesson 2			
Sequence of operations	15 mins	15 mins	15 mins	15 mins	15 mins	15 mins	15 mins	15 mins
Collect materials identified in cutting list, collect marking-out tools								
Mark out for holes and counterbores								
Checkpoint: check marking out with drawing								
Collect Forstner bits and drill bits								
Set up pillar drill and drill as required (PPE)								
Clean down pillar drill, return all Forstner bits and drill bits								
Checkpoint: check quality and evaluate								
Mark out for general shape and chamfer								
Checkpoint: check with drawing								
Cutting and shaping; remove waste material using hand tools								
Set up linisher and smooth sides and chamfer (PPE)								
Clean down linisher								
Checkpoint: check quality and evaluate								

Figure 3.63 An updated Gantt chart

The original Gantt chart (Figure 3.28, on page 191) demonstrated one-hour lessons broken down into four 15-minute sessions, with planned linear activity indicated by a blue line, checkpoints by a red diamond and evaluation by a green line. The additional brown markers in Figure 3.63 indicate the actual time taken, and they demonstrate a loss of around 15 minutes over two hours. While the chart represents reasonably accurate time management for the one-off activity, if this were a commercial enterprise making 1000 products, the loss of 15 minutes on each product would be very expensive for the business due to labour and energy costs, for example.

When evaluating the production plan, it is helpful to look closely stage by stage, tool by tool, process by process. Again, updating planning documents as work is ongoing, or making notes in a notebook or log book, will help when it comes to the summative evaluation. For example, when a tool or machine is not available, the engineer could update their plan with a solution to keep the project on track. Following on, they could comment on the effectiveness of the solution in their notebook.

Evaluation of outcome

Evaluation of a completed outcome should be based largely on what was set out to be achieved in the initial brief (type of product required) and specification (characteristics and performance requirements).

When comparing the outcome against the specification criteria, the engineer is seeking to provide an objective evaluation. Some thought must be given to how each specification criterion will be tested so that reliable data is achieved, data that would not change if the same tests were performed by another person; it is important that evaluation comments are factual and relevant, and that they are fully justified with sufficient detail to support the evaluation judgement.

Due to often conflicting demands for what is wanted and what is possible, final outcomes do not always meet all of the specification criteria, so it is worth checking which of the criteria are critical, which are **desirable** and which are **non-essential**. Clearly, all of the critical criteria issues should be available to be evaluated in detail because they were identified as being essential to the product, but desirable and non-essential issues may not have been met in the final outcome because they are not as important.

> ## Key terms
>
> **Critical criteria** Criteria that must be achieved.
>
> **Desirable criteria** Criteria that should be achieved.
>
> **Non-essential criteria** Criteria that could be achieved but are not required.

For example, specification criteria for a bicycle light indicate that it must provide appropriate lighting from high-output LEDs, that it should be able to be attached to handlebars, and that it could be available in a range of colours. Therefore, the bicycle light should be evaluated to see if it meets the criteria.

In order to provide factual, relevant comments, the bicycle light should be inspected and tested to answer the following questions:

- Have high-output LEDs been used in the construction?
- Do the LEDs provide appropriate lighting?
- Can the light be attached to the handlebars of a bicycle?
- Is the bicycle light available in a range of colours?

These are relevant and reasonably easy to answer with facts. For the first point (a critical [must] criterion), a visual inspection of the LEDs for code markings or a check of the documentation that came in the box may be sufficient to show if high-output components

have been used, and justification can be provided by explaining the level of scrutiny taken (how the information was gained).

The second point about providing appropriate lighting (again, a critical criterion) is a little unclear because it does not state when or in what conditions the lighting needs to be appropriate. Therefore, tests may be required to show if the bicycle light provides adequate lighting in a range of conditions. Justification may explain what was learned from the testing.

The third point about being able to attach the light to a bicycle's handlebars is a desirable (should) criterion. Once again, testing would prove if the bicycle light can be attached to handlebars. However, note that the criterion could still be met if the bicycle light could not be attached to the handlebars but it could be attached to the frame of a bicycle; justification would explain this.

The final criterion about availability in a range of colours is a non-essential (could) criterion, so if it is not available in a range of colours and only in a neutral grey colour, the criterion could be justified as being met.

Justification is an explanation of the evaluation judgement and could include some explanation of how the judgement was achieved.

While the specification criteria provide a starting point for evaluation, particularly the essential criteria, there are often further issues that should be considered, some of which are outlined in Table 3.25. Addressing each point in turn can result in a detailed evaluation of the prototype model and identify areas for improvement.

ACCESS FM and similar analytical tools can be used to provide a framework for evaluation.

Table 3.25 Further issues for evaluation

Issues for evaluation	Questions to ask
Features	How successful were the features and details (such as switches, lights, dials, inputs and outputs)?
Function	Does the prototype do what it is designed to do? How well does the prototype perform? Does the environment where the prototype is used affect how it performs?
Materials	How appropriate were the materials that were selected for the prototype when considering environment of use, weight, strength, serviceability, maintenance and cost?
Aesthetics	Is the prototype visually appealing? How do shape, colour, tone and texture add to or detract from the design?
Ergonomics	How easy is it for the average person to interact with the prototype (to use/operate/open/close/lift/place down)? Consider average measurements (such as height, hand or foot size) for both males and females, and how comfortable the prototype is to use.
Modelling and prototyping processes	How successful were the modelling and prototyping processes? Could other processes have ensured a better quality of prototype? What computer-controlled processes were used in modelling and prototyping?
Alternative manufacturing techniques	What other manufacturing techniques could be used to realise the prototype, such as computer-controlled, rapid-prototyping and automated processes?

Case study

When James Dyson was developing a design for a new vacuum cleaner, he produced more than 5000 prototypes before he came up with a final version. Every component would have been prototyped, evaluated, developed and evaluated again, until it provided exactly what Dyson wanted from it.

Identify potential improvements in the design

There are very few prototype products that do not have the potential of being improved. Prototype modelling is an opportunity to test designs and theories, to identify and correct issues that were not recognised during design stages; errors in a prototype design or even failure to function as intended do not necessarily indicate bad outcomes. In practice, prototypes do not always appear or function exactly as intended – they will have strengths and weaknesses.

Although the time Dyson took to perfect his design is not available for this unit of work, the principle of identifying where potential improvement could be achieved is the same and it is included in the assessment criteria.

Research the Dyson vacuum cleaner story.

Commercial manufacturers will often make many different prototypes, testing different manufacturing techniques to find the most suitable, a luxury the model-maker does not necessarily have. In schools and colleges, products will be prototyped from resources that are available at the time of modelling, so there will always be materials, tools, equipment and processes that could have been used instead. The model-maker should consider if by using different materials or manufacturing processes they could have improved their prototype design.

Thorough evaluation of a product against its specification criteria will help to identify its strengths and weaknesses; strengths are apparent where specification criteria have been met fully, weaknesses where criteria are only partially met, or not met at all. Weaknesses indicate areas that could lead to potential improvements. Once weaknesses have been identified, they should be explained in precise factual statements with some justification for where improvement could be achieved.

For example, a prototype litter-picker has been tested and a major fault was identified. The test is explained with justification for the suggested improvements to the design:

The litter-picker was designed to grip and hold litter up to the size of a large drinks bottle. It worked well when picking up general wrapper-type litter and empty drinks bottles but not when picking up and holding bottles or cans that were around half-full with a liquid. These slid out of the jaws of the litter-picker.

Improvement of the mechanism for gripping and holding could make the grip action more secure, and a non-slip surface added to the litter-picker's jaws could stop the bottles and cans slipping out of the jaws.

When evaluating a product's design and manufacture, it can be helpful to use questions that provide straightforward yes/no answers

Table 3.26 Evaluation-type questions that provide yes/no answers

Question	Answer
Does it (the prototype) do what it was designed to do?	Yes/No
Is its size appropriate?	Yes/No
Can it be used in a range of locations?	Yes/No
Will it meet the needs of a wide range of users?	Yes/No
Does it have visual appeal?	Yes/No
Is it a high-quality prototype model?	Yes/No
Can it be made at a low cost?	Yes/No
Will it be available in a range of colours?	Yes/No
Does it work from batteries and mains power?	Yes/No
Is it finished to a commercial standard?	Yes/No

(such as 'does it?', 'is it?', 'can it?', 'will it?'), as shown in Table 3.26. These are questions that link to specification criteria.

The yes/no answers can be followed up with questions such as:

- How well?
- Why not?

Answers to these questions should provide explanations and lead to suggestions for potential improvements.

Research tools

Another way of examining products is by using primary research tools, such as interviews, consumer trials, surveys and questionnaires, to gain the objective views of a sample user group. These can be helpful because they can provide qualitative objective data. Specification criteria can be turned into questions for an interview or written on a questionnaire with digital yes/no responses and, most importantly, with some space for responder feedback comments and judgements.

An alternative to yes/no responses could be feedback rating symbols to represent very good, satisfactory and poor, as shown in Figure 3.64. Again, you should leave space for feedback comments or judgements because these comments can provide the justification for potential improvements.

The results from primary research could reinforce your own views or they could point to issues that you had not previously identified or considered, and lead to areas for potential improvement.

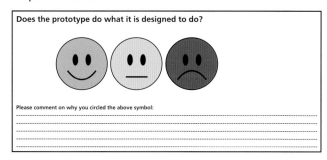

Figure 3.64 An example of a consumer feedback questionnaire with feedback rating symbols

The quality of a finished outcome can be affected by the resources available at the time of modelling (such as the materials, tools, equipment, processes, facilities). Where this is the case, you should identify unsatisfactory resources and suggest ones that could lead to potential improvements. When judging the quality of an outcome, it is extremely difficult to achieve commercially manufactured product quality in a non-commercial manufacturing facility. Therefore, like-for-like assessments (commercially manufactured vs school/college manufactured) are somewhat unrealistic, but they do provide areas for potential improvement.

Final outcome quality can also be affected by the skills set of the model-maker. Non-commercial manufacturing processes that replicate casting, moulding, forming, joining, and so on are very difficult to master and errors that result from the skills of the model-maker should be judged as you would for a weakness in a material or process. Evaluation is not there to judge the person, but to point to how the outcome could potentially be improved as a result of skills development.

Activity

Many commercially produced products are potentially improved within a year or two of release as a result of **market pull** and **technology push**.

● Identify two products that have been updated as a result of market pull and comment on their updates.
● Identify two products that have been updated as a result of technology push and comment on their updates.

Key terms

Market pull When a need for a product arises from consumer demand, which 'pulls' the development of the product.

Technology push When, as a result of research and development (R&D), new technology is created that can lead to the development of new products; products are 'pushed' into the market, with or without demand.

Test your knowledge

1 Why is high-quality evaluation essential for the success of a business?
2 What type of evaluation is based on personal opinions?
3 Specification criteria are the only criteria against which a product should be evaluated. True or false? Explain your answer.
4 When should planning documents be evaluated?
5 How can reviewing one's own performance in realising a design be helpful to an engineer?

Assignment practice

Marking criteria

Mark band 1: 1–2 marks	Mark band 2: 3–4 marks	Mark band 3: 5–6 marks
Produces a **basic** evaluation of the prototype outcome against the product specification.	Produces an **adequate** evaluation of the prototype outcome against the product specification.	Produces a **comprehensive** evaluation of the prototype outcome against the product specification.
Provides **limited** potential improvements. No justification is provided.	Provides **some** potential improvements, with justification.	Provides **detailed** potential improvements with justification

Top tips

When presenting the evaluation for assessment:

- Review the quality and accuracy of your planning documents; briefly explain ongoing updates and why they were necessary.
- Use the specification criteria as the starting point for evaluation of the prototype and then extend the evaluation to cover a wide range of issues.
- Evaluate with a critical eye and justify each statement.
- Use photographs to illustrate evaluation comments.
- Use primary research tools to generate qualitative data that can help to identify areas for potential improvements.

- Suggest potential improvements for the prototype and justify each suggestion.

Note: This unit does not require any of the suggested modifications to be made. It is important that each potential improvement is justified, with reasons to suggest how the modification could potentially improve the prototype. Individual components could be evaluated separately.

Model assignment

For the prototype gate hold-back catch produced for the previous model assignment:

- Produce a detailed evaluation of the outcome.
- Identify and justify potential improvements.

Example candidate work

Evaluation of my prototype against the specification

Evaluation against the specification:

The glue gun prototype's handle is thin and ergonomiclly shaped to fit the average hand sizes of adult men and women, allowing for a comfortable grip. The glue gun also has a quality 'solid' feel due to the thickness and denity of the MDF used to create the prototype. The design for the glue gun can either be injection moulded or vacuum formed due to their symmetrical designs, allowing for both sides to be easily screwed together. The metal stand of the glue gun is an appropriate enough length, width and shape to withstand the weight and size of the glue gun when used to balance it. When not in use, howveer, the stand is not symmetrical, so the glue gun is rested at an angle. The hole running horizontally from the back of the nozzle has an 18mm diameter, allowing for most standard sized glue sticks to fit, allowing for smooth and easy removal and insertion of a glue stick. The positioning of the trigger is in the centre of the handle, giving easy access to it no matter the hand size of the user. The bright purple colour of the gun and its black detailing (in the nozzle, trigger and logo) make the glue gun easily recognisable, with the spray paint itself providing enough protection and gloss that any further surface finishing would be required. Furthermore, the logo of the gun is exactly the same on both sides, rather than mirroring, which was an oversight made as I rushed to design them.

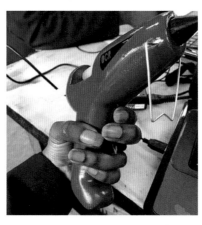

Following feedback from customers, OCR Product Development has designed a new **hot melt glue gun** to increase its range of workshop modelling tools. This product is aesthetically pleasing and has an ergonomic grip that allows greater comfort during use.

The product specification is shown below:

The glue gun must:
- allow both adult male and female users to grip and use the glue gun comfortably
- should be able to independently stand when not in use
- have a quality 'solid' feel
- be a 'two piece' design that could be injection moulded
- allow ease of glue stick installation/replacement
- be comfortable to hold and easy to use
- not require any surface finishing
- be readily identifiable.

Improvements I could make to my prototype

Although two layers of purple spray paint were distributed on the glue gun, there are still some scratches that could've been avoided if I had spray painted after I finished all the drilling.

The tip of the nozzle is scratchy and sharp to the touch, which can cause a hazard, potentially injuring the user or any nearby people. The nozzle should be rounded off at the tip so it is smooth with no sharp edges.

The trigger could have been made to be wider, as to be more ergonomic and provide more surface area for the user to place their forefinger. This would make the glue gun more comfortable to hold and easier to withstand extended periods of being handled.

Heat grills are a possible improvement which would increase the safety of the glue gun. Although included in the CAD design, such improvement was unable to be produced due to time constraints, however this would be beneficial to a functioning glue gun in order to provide ventilation so the heating mechanics wouldn't overheat.

The stand itself is lopsided, due to the errors I made as I moulded it by hand. To fix it I'd start completely afresh with a new piece of aluminium rod, cut to size. Then I would be very cautious and slow when using a vice and pliers to mould it, as to make all bends straight and symmetrical.

When 2D designing my logo stickers, I would make sure that they mirror each other, in order to have the desired symmetrical look of the glue gun. I would also stick them more central to the glue gun, as to make it more recognisable.

A more gender neutral colour could be chosen for my glue gun, as to make it appealing to both men and women. Colours such as white, silver and gold would be suitable due to the connotations of high quality that they bring to products.

When repainting the glue gun, I would make sure to do the process much slower and with less force on the nozzle of the spray can. This would help even out any run off.

Instead of inserting my trigger by drilling holes, I'd use MDF filler to fill in the holes. Then I'd file down the trigger until the tailpieces are gone, so I could use glue from a hot glue gun to stick it onto the glue gun. I would then use a scribe to remove any excess glue from the sides of the trigger in order to make the trigger more secure and easier to assemble.

Evaluation of my performance

My strengths and weaknesses:

Glueing the two MDF blocks was successful due to the keying done prior, which allowed for the glue to effectively bond the pieces together, a useful idea. Filing down the glue gun after cutting the outline was a step that I was particularly good at due to my patience and use of feedback which allowed me to gain the ergonomic effect I wanted.

Sanding down was also a strength as I used methods of sanding that were very efficient to quickly remove and smoothen pieces of MDF, to get the finish I required without harming my fingertips through abrasions. When drilling, I was very cautious and precise, making sure that the holes were symmetrical with my great attention to detail.

However, when 2D designing the logo, I was too headfast, leading to mistakes that could have been avoided if I visualised what I was making, to ensure that it was appropriate and functional.

Moreover, I was not careful enough when transporting the prototype around, resulting in it receiving scratches on its paint job, which eventually required a repaint. When I was making the stand, I lacked the arm strength to cut the aluminium rod by myself, as well as to mould the rod quickly and effectively, so the rod would still be straight.

My spray paint was done effectively the first time around, yet in the second round, I was too absent-minded when spraying, allowing for the paint to build up in one concentrated area due to my lack of focus.

Evaluation of my performance:

After careful appraisal of my performance, I would consider it to be a success. My filing was done well due to experience from previous engineering modules, matched with the feedback received from other pupils and teachers on the ergonomics, I was able to change the shape of the handle to be ergonomic and fit comfortably in a range of hand sizes. Moreover, the holes drilled for the trigger and stand were precice and symmetrical, due to my cautious approach to drilling. However, my spray painting was often rushed at times, which lead to slight paint run off. In addition, my stand was also slightly wonkier than I wanted, however it still managed to keep the glue gun off the workbench when used so it was still functional.

Evaluation of my prototype against my production plan

Evaluation against the product plan:

My product encountered some problems that led it astray from the plan, predominantly in the earlier stages. When glueing together the MDF blocks, too much glue was placed, resulting in the excess glue spilling out over the sides. However, the problem was not too significant as the outline of the glue gun was cut out, getting rid of the sides of the original blocks of MDF. In addition, I had to go over the spray paint I did after drilling holes in the handle, as at my workbench, the gun had its paint scratched, requiring a second coat, which resulted in some of the spray paint run-off drying due to the excess spray paint. Moreover, I only used the 180 and 600 grit papers, rather than using all grit papers in ascending order.

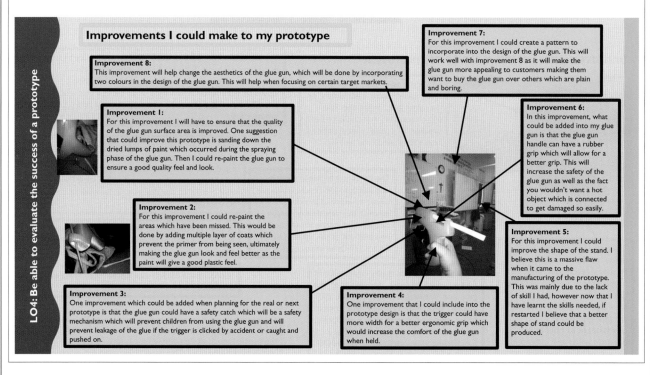

LO4: Be able to evaluate the success of a prototype

Improvements I could make to my prototype

Improvement 8:
This improvement will help change the aesthetics of the glue gun, which will be done by incorporating two colours in the design of the glue gun. This will help when focusing on certain target markets.

Improvement 7:
For this improvement I could create a pattern to incorporate into the design of the glue gun. This will work well with improvement 8 as it will make the glue gun more appealing to customers making them want to buy the glue gun over others which are plain and boring.

Improvement 1:
For this improvement I will have to ensure that the quality of the glue gun surface area is improved. One suggestion that could improve this prototype is sanding down the dried lumps of paint which occurred during the spraying phase of the glue gun. Then I could re-paint the glue gun to ensure a good quality feel and look.

Improvement 6:
In this improvement, what could be added into my glue gun is that the glue gun handle can have a rubber grip which will allow for a better grip. This will increase the safety of the glue gun as well as the fact you wouldn't want a hot object which is connected to get damaged so easily.

Improvement 2:
For this improvement I could re-paint the areas which have been missed. This would be done by adding multiple layer of coats which prevent the primer from being seen, ultimately making the glue gun look and feel better as the paint will give a good plastic feel.

Improvement 5:
For this improvement I could improve the shape of the stand. I believe this is a massive flaw when it came to the manufacturing of the prototype. This was mainly due to the lack of skill I had, however now that I have learnt the skills needed, if restarted I believe that a better shape of stand could be produced.

Improvement 3:
One improvement which could be added when planning for the real or next prototype is that the glue gun could have a safety catch which will be a safety mechanism which will prevent children from using the glue gun and will prevent leakage of the glue if the trigger is clicked by accident or caught and pushed on.

Improvement 4:
One improvement that I could include into the prototype design is that the trigger could have more width for a better ergonomic grip which would increase the comfort of the glue gun when held.

These examples demonstrate a systematic approach to evaluation of the prototype model. Specification criteria are reproduced and the prototype is evaluated against them. Potential improvements are identified, with justifications for each.

Research

For further information on writing an evaluation, look at the following websites:

Writing evaluations: **https://technologystudent.com/designpro/eval1.htm**

Evaluating against the specification: go to the BBC Bitesize website and search for 'Evaluating – Eduqas'.

Synoptic links

Unit R040 allows you to apply the key knowledge, skills and understanding that you learned in Unit R038, particularly with reference to:

- using different research methods to research existing products, including through safe disassembly
- applying techniques, such as ACCESS FM and ranking matrices, to review and compare products
- how to read engineering drawings and use CAD to virtually simulate design ideas
- how to create and evaluate physical prototypes of products as part of the design cycle.

You also develop skills that you have learnt in Unit R039 about engineering drawings and using CAD software to further apply these skills in Unit R040.

Glossary

3D printing Production of a 3D physical component from a computer-aided design model, by adding material layer by layer.

5WH method Questions that can be used to interrogate a design (Who, What, Where, When, Why and How).

6Rs Areas to be considered when assessing the sustainability of a product: recycle, reuse, repair, refuse, reduce and rethink.

Abrasion Process of wearing something away.

ACCESS FM Acronym for Aesthetics, Cost, Customer, Environment, Size, Safety, Function and Materials; a product analysis tool.

Acrylonitrile butadiene styrene (ABS) Tough, rigid thermoforming polymer with high-impact strength.

Additive manufacturing Technologies that produce 3D components and products from CAD data by adding material layer by layer (as opposed to subtractive manufacturing processes that remove material from a larger block).

Aerosol Mixture of liquid droplets or minute solids in air or another gas – for example, spray adhesive or spray paint.

Aesthetics How well a product appeals to the senses.

Allen key Hexagonal-shaped tool that fits in a socket-head screw.

Alloy A mixture of two or more metals.

Animation Videos produced by CAD software for visualising products.

Annotations Notes or comments added to a sketch or diagram that provide explanation and give meaning.

Anodised Coated with a protective oxide layer using an electroplating process.

Anthropometrics Study of the measurements of the human body.

Assembly Putting together components to make a completed product (if the resulting product is to be incorporated into a larger product, it is called sub-assembly).

Assembly drawing Drawing showing how separate components fit together.

Assembly instructions Instructions included with a product with drawings and text on how to assemble the product.

Autoclave A sealed chamber in which products can be heated and pressurised under controlled conditions.

Automation Using computer technology to operate equipment, rather than humans.

Axes (singular axis) Reference lines for the measurement of coordinates – usually x, y and z in 3D space.

Batch production Method used in manufacturing where products are made in a specific amount (a batch) within a specific time frame.

Bespoke Made specifically for a particular customer or user.

Bill of materials Another name for a parts list.

Bioplastics Biodegradable plastic materials produced from renewable sources such as corn starch, vegetable fats and oils.

Block A rigid piece of material that is supplied with relatively flat surfaces.

Block diagram Diagram that shows in schematic form the general arrangement of the parts or components of a complex system or process, such as an industrial apparatus or an electronic circuit.

Breadboarding The construction of an electronic circuit on a board (solder or solder-free) using jumper wires to transfer voltage around the breadboard.

Break-even point The point where enough products have been sold to cover the development costs.

Budget Amount of money allocated by a client or company to develop a product.

Cabinet view Technique used in oblique drawing where the distinguishing face is drawn original size but the other sides along the lines of sight are drawn half size.

Carcase A hollow structural enclosure for containing an object, such as a motor, electronic circuit or mechanism.

Casting Process where solid material is heated until it turns into a liquid and is then poured into a mould; once cooled, the material will have taken the shape of the mould.

Cavalier view Technique used in oblique drawing where all sides of the object are drawn original size.

Centre lines Lines drawn to indicate the exact centre of a part; always drawn using a series of shorter and longer dashes (or longer dashes and dots).

Chamfer A transitional edge (or cut-away) between two faces of an object.

Characteristics A noticeable or typical feature.

Chuck Specialised clamp that holds material so it can be turned on a lathe.

Circuit diagram Graphical representations of electric circuits, where separate electrical components are connected to one another.

Client Person, group of people or company that has commissioned the development of a new product.

Competitive advantage When a business is in a favourable position compared to other businesses because it has products, technology or market share that others do not have.

Components Parts or elements of a larger assembly.

Composite materials Materials made up of two or more different materials, combining their properties to create a new, improved product.

Compound shape A shape made from a number of different shapes or elements put together.

Compressive strength Strength of a material under load (when the load is 'compressing' the object).

Computational fluid dynamics (CFD) Method of simulation undertaken in a software package that analyses how a gas or liquid flows through or around components and products.

Computer-aided design (CAD) Using computer software to develop designs for new products or components.

Computer-aided manufacture (CAM) Using computer software to control machine tools.

Computer numerical control (CNC) machining Using computer-controlled machine tools to remove material from a workpiece to create components.

Concept prototype Prototype model that appears as a final product might but often does not function as a fully working product would.

Concept sketching Producing drawings quickly and often by hand in order to explore initial design ideas.

Consumables Resources that assist manufacture and are used up during the process – for example, oil and lubricant used in machines.

Control measures Actions taken to reduce the risk (likelihood of a hazard causing harm), such as removing the hazard, taking extra care, guarding or wearing protective equipment.

Control of Substances Hazardous to Health (COSHH) Regulations 2002 Legislation that provides a framework to protect people in the workplace against health and safety risks from being exposed to hazardous substances.

Correlation Relationship between two variables. Can be positive, negative or none.

Corrosion Gradual destruction of materials due to a reaction with their environment or chemicals.

Countersink A conical hole cut into an object so that a bolt or screw can be sunk below the surface.

Critical criteria Criteria that must be achieved.

Cutting and wasting Removal of unwanted material.

Cutting list List of the materials required at their prefabrication size so that modelling or making can commence.

Cutting plane line Line showing where an imaginary cut is made through an object to expose a sectional view.

Demographic Used to describe the numbers and characteristics of people who form a particular group.

Design cycle A set of processes, split into four phases, that designers follow to ensure efficient and effective product development.

Design for disassembly Features added to a design that allow it to be easily taken apart for cleaning, maintenance or disposal.

Design for manufacturing and assembly (DFMA) Process where a product is designed for efficient manufacture and assembly.

Design for the circular economy Economy model based on economic growth without consumption of finite resources.

Design process A series of stages that designers and engineers use in creating functional products.

Design specification Detailed document that defines all the criteria required for a new product.

Design strategy A series of stages that are part of the design process.

Desirable criteria Criteria that should be achieved.

Deteriorate Get worse with age or fall apart.

Dimensions Numerical values added to engineering drawings to communicate the sizes of key features (measurements are usually in millimetres).

Disassembly Taking something apart (for example, a product or piece of machinery).

Disposal Stage in a product's life cycle when it is no longer useful and must be thrown away or recycled.

Distinguishing face Face or side of an object with features of most importance.

Draft angle Sloping face on the wall of a component, set at a specific angle so that it can be removed from a mould.

Drawing border Line around a drawing that delineates the design area and often provides grid references to easily identify areas of the drawing.

Drawing (or work) planes Computer-based representations of 3D space.

Drill bit A cutting tool used to create holes by removing material.

Ductility The ability of a material to be stretched under load without breaking.

Durability Ability of a material to withstand wear, pressure or damage.

Emerging market Either a part of the world or a group of consumers that has been identified as potential future customers based on developing trends and behaviours.

Engineering drawing Type of technical drawing that details the geometry, dimensions and features of a component or product.

Ergonomic design A process for designing products using anthropometric data so that they perfectly fit the people who use them.

Ergonomics Science of designing products so that users can interact with them as efficiently and comfortably as possible.

Error proofing Integration of a mechanism or device into a product or process that stops it being misused, prevents it being assembled in the wrong way or protects the user.

Exploded assembly A drawing where the components of a product are drawn slightly separated from each other and suspended in space to show their relationship or the order of assembly; also known as an exploded view.

Exploded view drawing Drawing that shows the component parts that make up the product separated and moved outwards.

Extrude tool CAD tool used to create a 3D solid object by stretching a 2D profile sketch along an axis.

Fail-safe mechanisms Design features integrated into a product to protect the user from harm in the event of a fault or misuse.

Ferrous material Material that contains iron.

Final renders Realistic images of computer-generated models to show what a finished product looks like without the need to produce a physical model.

Finishing operations Operations carried out on a component to make physical corrections (for example, removing sharp edges) or to add a surface finish (for example, painting).

Finite element analysis (FEA) Method of simulation undertaken in a software package that analyses how a component is affected by applied forces or stresses.

Fixed-position power tools Motor-driven tools that cannot practically be moved and are usually bolted down – for example, a pillar drill.

Flowchart Block diagram that shows how various processes are linked together to achieve a specific outcome.

Focus group Group of people invited to discuss their views and opinions on their wants, needs and preferences for a new product.

Former Device over or around which materials can be formed.

Fully defined Description of a correctly dimensioned profile.

Functional prototype Prototype model that has the appearance and performance of a final product.

Functionality The purpose for which something is designed or expected to fulfil.

Gantt chart Planning chart that has a sequence of operations plotted against time.

Geometry Shape of an object.

Grain Natural alignment of fibres seen in a cut surface of wood.

Hatched lines Diagonal parallel lines which show where an object has been cut.

Hazard Something that presents a danger, either to physical health such as electricity or an open drawer, or to mental health.

High-impact polystyrene (HIPS) Rigid thermoforming polymer with high-impact strength.

Hole A hollow place or opening in a solid body or surface.

Incinerated When something is burned to dispose of it.

Inclusive design A design process where a product is optimised for a specific user with specific needs.

Interface Means by which a user interacts with software.

Isometric drawing 3D pictorial drawing that focuses on the edge on an object and uses an angle of 30° to the horizontal.

Iterative design A circular design process that models, evaluates and improves designs based on the results of testing.

Jigs and fixtures Tools used in manufacturing to ensure components are placed or held accurately so that they can be replicated consistently.

Knurling A manufacturing process, typically done using a lathe, where a pattern of straight, angled or crossed lines is rolled into the material.

Labels Text added to a sketch that points out details, features and characteristics – for example, switch, LED and battery holder.

Labour costs Cost associated with employees in a business, including wages, taxes and additional benefits.

Labour intensive Needing a large amount of effort from a workforce in relation to the amount of output produced.

Laminate Layers of material that have been compressed and bonded together by an adhesive.

Leader line A line that points to a significant feature on a drawing.

Legislation Laws proposed by the government and made official by Acts of Parliament.

Life cycle analysis (LCA) Technique used to evaluate the impact of a product on the environment at all stages of its life, from its creation to the point it is disposed of.

Linear design A design process where the stages are carried out one after another, often without turning to any of the previous stages.

Linear economy Economy model based on the extraction of natural resources for products that will eventually end up as waste and potential pollutants.

Linear measurement Measurement indicated in a straight line.

Linear movement Movement in a straight line, forwards and backwards.

Lines of sight Parallel angled lines drawn to help construct 3D drawings.

Local exhaust ventilation (LEV) Control system designed to reduce exposure to airborne pollutants and contaminants, such as dust, fumes, gas and vapour, by taking them out of the workplace.

Log book Ongoing record of questions, decisions and solutions that can be used as evidence of systematic activity and review.

Low carbon steel A low carbon ferrous material (contains iron) that consists of less than 0.3 per cent carbon; also known as mild steel.

Malleability The ability of a material to be shaped or deformed by compressive forces (such as hammering or pressing).

Mandatory signs Safety signs that indicate an action must be taken.

Manufacturing plans Detailed documents that set out the material requirement, production quantity, production setup and process, and timescales for making a product.

Manufacturing process Stages through which raw materials go in order to be transformed into a product.

Market pull When a need for a product arises from consumer demand, which 'pulls' the development of the product.

Market research Process of gathering information about the needs and preferences of potential customers.

Mass production The production of a large quantity of a standardised product or component, often using automated production processes; also known as flow/continuous production.

Mate tool CAD tool used to align parts accurately and fit them together (sometimes called mate constraint).

Mechanism Set of components that work together in a product to carry out a function.

Micrometer A device that measures small distances or thicknesses between its two faces, one of which can be moved away from or towards the other by turning a screw with a fine thread.

Milling A machining operation designed to cut or shape material using a rotating cutting tool.

Morse taper A machine taper in the spindle of a machine tool or power tool.

Motion study A simulation tool in CAD software allowing the study of how parts and components move in relation to each other.

Mould Device for producing a 3D form in the shape of the desired outcome.

Multi-functional product Single product that can carry out the tasks of multiple products.

Multimeter An instrument designed to measure electric current, voltage and usually resistance, typically over several ranges of value.

Needs Aspects of a design that are considered to be critical to the future outcome.

Non-destructive testing (NDT) Where components are tested without the need to damage or destroy them.

Non-essential criteria Criteria that could be achieved but are not required.

Objective evaluation Appraisal that is based on fact, is reliable and could be repeated if performed by another person.

Oblique drawing 3D pictorial drawing that focuses on the face of an object and uses an angle of 45° to the horizontal.

Obsolete When a product becomes outdated or unserviceable and parts are no longer available to repair them.

One-off production Manufacturing products one at a time.

Origin Point where axes meet.

Orthographic drawing Drawing that represents a 3D object by using several 2D views (or projections) of it.

Overheads Expenses that need to be paid by the business, not including labour or materials, such as rent and utilities.

Part warping When a moulded component deforms from its desired shape because parts of the material cool and shrink at different speeds.

Parts list List of all the individual components required to make a working product, which usually includes component names, quantities, materials, costs and suppliers.

Pattern Reusable, shaped piece of robust material that can be used to guide a manual machine-tool operation.

Perceived obsolescence A marketing strategy that encourages consumers to upgrade to the latest model.

Peripheral devices Devices that connect to a host product, such as a watch, speaker or earphones (peripherals) connecting to a mobile phone (host).

Personal protective equipment (PPE) Equipment designed to protect the user against risks to their health or safety.

Physical prototype 3D physical model of a component or product that allows the designer, client or user to interact with it.

Physical testing Testing undertaken on an actual, physical prototype to see how it reacts to real-world conditions or forces.

Pictorial drawing View of an object as it would be seen from a certain direction or point of view.

Pilot hole A small hole drilled in material.

Planned obsolescence A strategy of planning a deliberately shorter lifespan for a product.

Planning table Formal planning sheet that records all the relevant information required for modelling and making.

Polygon A plane (flat) figure with at least three straight sides and angles, and typically five or more.

Polylactic acid (PLA) Natural polymer made from corn starch or sugar cane.

Polymer Material made from chains of a repeating chemical part called a monomer; examples include thermoplastics, thermoset plastics, expanded foam plastics and smart plastics.

Portable power tools Motor-driven tools that can be carried around and used in different locations – for example, a hand drill or a palm sanding machine powered by battery or by mains electricity.

Pre-manufactured components Components or sub-assemblies manufactured separately from the whole product, sometimes by an external supplier, that are then assembled into the final product.

Primary raw materials Naturally occurring substances extracted from the earth.

Primary research Gathering original information first hand – for example, carrying out interviews, experiments, questionnaires and surveys.

Process planning Where all activities required to complete the development of a new product are defined with timescales assigned, to ensure the product can be delivered on time and within budget.

Product disassembly Taking a product apart to look at the materials, parts, components and fixings that have been used.

Product performance How well a product can carry out its task.

Production methods How things are made; there are three main types: one-off production, batch production and mass production.

Projection lines Lines used to extend existing lines on a drawing and used to help create new geometry.

Projections 2D views of an object used to represent it in 3D.

Prosecuted Officially accused in court of breaking a law.

Prototype Model of a component or product created either in a software package or physically that can be used to test or check the design.

Qualitative data Data based on descriptions and observations, which cannot be counted or measured.

Quality assurance Monitoring and review of all processes required to produce a desired outcome at a desired standard.

Quality control testing Pre-arranged testing performed at specific stages in the process to ensure that outcomes are as required.

Quantitative data Data based on numbers and quantities, which can be counted or measured.

Ranking matrix A table with numbers assigned to rate product features and used to compare products.

Ratio Comparison of two or more numbers that indicates their sizes in relation to each other.

Raw materials Extracted materials that will be processed and then used to produce components.

Reference geometry Defines the shape or form of a surface or a solid, and includes items such as planes, axes, coordinate systems and points.

Remanufactured Rebuilding of a product to its original specification through repair or the use of new parts.

Rendered imagery Photorealistic computer-generated images of products.

Rendering tools CAD tools that generate a realistic 3D image by adding features such as lighting, shade, reflections, tone and texture.

Research and development (R&D) Often the first stage in a development process, where companies carry out research activities to innovate and support the introduction of new products.

Resistance to corrosion Ability of a material to resist deterioration caused by reactions to its surrounding environment.

Reuse Practice of using something again for its original purpose or to fulfil another function.

Reverse engineering Taking apart and analysing a product's construction or composition in order to produce something similar.

Revolve tool CAD tool used to create a 3D solid object by rotating a 2D profile around a centre line.

Risk The chance or likelihood that someone could be harmed by a hazard and how serious the harm could be.

Risk assessment Process of identifying, analysing and evaluating hazards and their associated risks, and seeing if the risks are acceptable or can be reduced.

Robust Strong, hard-wearing, less likely to break.

Rotational movement Movement around an axis, such as hinge movement.

Router High-speed hand-held or table-mounted rotary cutter.

Safety data sheets (SDSs) Written documents that provide information and procedures for the safe handling and use of chemicals.

Sample size Number of people included in a sample; it needs to be large and diverse enough to ensure the data represents the population.

Scale Amount by which a drawing is reduced or enlarged from the size of the actual object, shown as a ratio.

Scale of production Number of products to be produced to meet demand or by a certain production process – for example, one-off, batch or mass production.

Secondary research Gathering information from sources that already exist – for example, using books, newspapers, magazines and the internet.

Sectional view View used on an engineering drawing to show a 'cut-away' cross-section of an object so that the internal features can be detailed.

Service manual Document that provides step-by-step guidance on how to disassemble, maintain and reassemble a product safely.

Sheet A flat material up to approximately 10 mm in thickness.

Shell tool CAD tool used to hollow out a solid part.

Simulation Where a computer-generated model of a component or product is exposed to virtual conditions that represent real-world scenarios to analyse how the product or component reacts.

Simulation tools CAD tools that allow a model to be analysed virtually using engineering techniques and scientific calculations.

Smart materials Materials that change in response to stimuli in the environment.

Solid modelling Computer representation of a 3D solid object that can be used in design and simulation.

Spline Smooth curve through a set of points.

Sprue Passage created to pour molten material into a mould (the excess material that needs to be removed as a result of this process is also called a sprue).

Standard An agreed way of doing something, such as making a product, managing a process or delivering a service.

Standard components Individual components made in a large quantity (often by mass production) to the same specification.

Standard forms Made available in large quantity (often by mass production) to the same specification.

Standard stock sizes Materials and components that are readily available in a range of sizes (such as sheet material at 60 cm x 60 cm, 1.22 m x 2.44 m, and so on; nuts and bolts at the following standard sizes: 4 mm, 5 mm, 6 mm, 8 mm, 10 mm, 12 mm, and so on).

Sub-assemblies Units of assembled components designed to be incorporated with other units and components into a larger manufactured product.

Subjective evaluation Appraisal based on personal views, which may include bias.

Summative evaluation Appraisal at the end of a series of processes.

Supply and demand Relationship between the quantity of products a business has available to sell and the amount consumers want to buy.

Supply chain Network of businesses that supply materials, components or services needed for the manufacture of a product.

Surface finish Nature of a surface, defined in terms of its roughness, lay (surface pattern) and waviness (irregularities in the surface).

Survey Tool used to gather data from particular groups of people that will inform the direction of a design.

Sustainability Meeting current needs without preventing future generations from meeting their needs.

Sustainable When something is used in a way that ensures it does not run out.

Sustainable design A design process where the designer attempts to reduce negative impacts of a product on the environment.

Target cost How much a company wants to sell a product for when it is put on sale.

Target market Group of people at whom the product being developed is aimed.

Technology push When, as a result of research and development (R&D), new technology is created that can lead to the development of new products; products are 'pushed' into the market, with or without demand.

Template Shaped piece of (usually) paper or card that can be used to test a profile or to mark around.

Tensile strength Strength of a material when it is stretched or pulled.

Thermoforming polymers Polymers that become pliable when heated, such that they can be shaped and formed, and harden when cooled.

Test of function Checking that the product works or operates in a proper or particular way.

Test of proportions Checking that the relationship between the size of different parts of a product are correct or attractive.

Test of scale (product) Checking that the overall dimensions of the product are correct or attractive.

Third angle orthographic drawing Orthographic drawing of an object which usually shows the front, right-hand side and top views.

Title block Box, usually included in the bottom right corner of a drawing, that includes important information to enable the drawing to be interpreted, identified and archived.

Tolerance Amount of variation allowed in a given dimension.

Tooling Manufacturing equipment needed to produce a component, such as cutting tools, dies, gauges, moulds or patterns.

Tooling costs Cost of moulds, cutting tools, jigs or fixtures required to make a product.

Toughness The ability of a material to resist impact or shock loads (such as press-forming a car body panel).

Trend Pattern of change that can be used to predict how the demands of a market are developing.

Turning A machining operation that generates cylindrical and rounded forms with a stationary tool.

Unibody The frame is integrated into the body construction; every panel provides part of the structural design.

Uniform thickness Consistent thickness throughout the whole component.

User Person or people who will use the final product.

User-centred design A design process where the designer focuses on the user and their needs in each step of the design process.

User testing Evaluating a product by testing it with representative users; sometimes called usability testing.

Veneer A thin sheet of wood that has been sliced off a tree trunk or log. Often used for decorative finishes.

Vernier caliper A measuring device that consists of a main scale with a fixed jaw and a sliding jaw with an attached measuring scale called a vernier.

Virtual prototype Model of a component or product produced in a software package that can be tested or used in simulations without the need to produce an actual model.

Virtual testing Testing undertaken on computer-generated representations of components or products to see how they react to real-world conditions or forces.

Visualisation Representation of an object.

Wants Desirable aspects of a design that are not considered to be critical to the future outcome.

Wiring diagram A simple visual representation of the physical connections and physical layout of an electrical system or circuit.

Wood veneer Very thin slice of wood up to approximately 3 mm thick.

Working drawing A scale drawing which serves as a guide for the manufacture of a product.

Working environment Place where a product will be used or situated during operation.

Yoke and plug Two parts of a mould that can be used to shape pliable materials.

Index